D0114032

THE SECRET LIFE OF NUMBERS

THE SECRET LIFE OF NUMBERS

50 Easy Pieces on How Mathematicians Work and Think

George G. Szpiro

Joseph Henry Press
Washington, D.C.

Joseph Henry Press • 500 Fifth Street, NW • Washington, DC 20001

The Joseph Henry Press, an imprint of the National Academies Press, was created with the goal of making books on science, technology, and health more widely available to professionals and the public. Joseph Henry was one of the founders of the National Academy of Sciences and a leader in early American science.

Library of Congress Cataloging-in-Publication Data

Szpiro, George G.
The secret life of numbers : 50 easy pieces on how mathematicians work and think / George G. Szpiro.
p. cm.
Includes bibliographical references and index.
ISBN 0-309-09658-8 (cloth : alk. paper) 1. Mathematics—History. I. Title.
QA21.S995 2006
510—dc22

2005030601

Cover design by Michele de la Menardiere

Translated by Eva Burke, London

Printed in the United States of America

Dedicated to my parents
Marta and Benno Szpiro

my wife
Fortunée

and my children
Sarit, Noam, and Noga

Contents

PART V
CONCRETE AND ABSTRACT MATTERS

PART VI
INTERDISCIPLINARY POTPOURRI

Preface

Whenever a socialite shows off his flair at a cocktail party by reciting a stanza from an obscure poem, he is considered well-read and full of wit. Not much ado can be made with the recitation of a mathematical formula, however. At most, one may expect a few pitying glances and the title "party's most nerdy guest." To the concurring nods of the cocktail crowd, most bystanders will admit that they are no good at math, never have been, and never will be.

Actually, this is quite astonishing. Imagine your lawyer telling you that he is no good at spelling, your dentist proudly proclaiming that she speaks no foreign language, and your financial advisor admitting with glee that he always mixes up Voltaire and Molière. With ample reason one would consider such people as ignorant. Not so with mathematics. Shortcomings in this intellectual discipline are met with understanding by everyone.

I have set myself the task of trying to remedy this state of affairs somewhat. The present book contains articles that I wrote on mathematics during the past three years for the Swiss daily newspaper *Neue Zürcher Zeitung* and its Sunday edition *NZZ am Sonntag*. It was, and is, my wish to give readers an understanding not only of the importance but also of the beauty and elegance of the subject. Anecdotes and biographical details of the oftentimes quirky actors are not neglected, but, whenever possible, I give an idea of the theories and proofs. The complexity of mathematics should neither be hidden nor overrated.

Neither this book nor, indeed, my career as a mathematics journalist evolved in a linear fashion. After studies of mathematics and physics at the Swiss Federal Institute of Technology in Zurich and a few career changes, I became the Jerusalem correspondent for the *Neue Zürcher Zeitung*. My job was to report about the goings-on in the Middle East. But my initial love for mathematics never waned, and when a conference about symmetry was to be held in Haifa, I convinced my editor to send me to this city in northern Israel in order to cover the gathering. It turned out to be one of the best assignments I ever did for the paper. (It was nearly as good as the cruise on a luxury liner down the Danube to Budapest, but that is another story.) From then on I wrote, on and off, about mathematical themes.

In March 2002 I had the opportunity to make use of my mathematical interests in a more regular fashion. The *NZZ am Sonntag* launched the monthly feature "George Szpiro's little multiplication table." I soon found out the hard way that the reception by the readers was better than expected: The incorrect birth date of a mathematician in one of the early columns led to nearly two dozen readers' letters ranging in tone from the ironic to the angry. A year later I received a special honor when the Swiss Academy of Sciences awarded the column its Media Prize for 2003. In December 2005, at a ceremony at the Royal Society in London, I was named a finalist for the European Commissionís Descartes Prize for Science Communication.

I would like to thank my editors in Zurich—Kathrin Meier-Rust, Andreas Hirstein, Christian Speicher, and Stefan Betschon—for their patient and knowledgeable editorial work, my sister Eva Burke in London for diligently translating the articles, and Jeffrey Robbins of the Joseph Henry Press in Washington, D.C,. for turning the manuscript into what I hope has become an enjoyable book on a subject commonly thought of as dry as a bone.

George G. Szpiro
Jerusalem, Spring 2006

I

Historical Tidbits

1

Lopping Leap Years

In early 2004 a phenomenon occurred that happens only about four times in the course of a century: There were five Sundays during the month of February. Such an event can only be witnessed every seventh leap year, that is, once in 28 years. The last time this happened was in 1976; the next time will be in 2032.

A lot is odd about leap years. Astronomers have observed that the time between two spring equinoxes is 365 days, 5 hours, 48 minutes, and 46 seconds, or 365.242199 days, which in turn equals nearly, but not exactly, 365.25 days. As an approximation this is good enough, though, and in the middle of the 1st century Julius Caesar introduced the calendar that would henceforth carry his name. Every three years, with 365 days each, would be followed by a leap year, which would include an additional day. For the following millennium and a half, the years thus had an average length of 365.25 days.

But toward the end of the 16th century the gentlemen of the Catholic Church were no longer prepared to put up with an annual error of 11 minutes, 14 seconds. Consultants to the Vatican had calculated that within 1,000 years the annual mistake would accumulate to a difference of eight entire days. Thus, in a mere 12,000 years, they pointed out, Christmas would fall in the autumn season and Easter would have to be celebrated in January. Since the Church plans for the long term, such inaccuracies are unacceptable.

Pope Gregory XIII (1502–1585) had a long think about all this and eventually arrived at the conclusion that the Julian year—the year according to Julius Caesar—was simply too long. To compensate for the inaccuracy, the Pope decided to adjust the calendar by leaping over some of the leap years: for each 25th leap year the leap day, originally added by Caesar, would be canceled. Thus, the month of February of the last year of each century—that is, each year that is divisible by 100—would only have 28 days, even though it should have been a leap year. The years whose leap days were being lopped off could thus be renamed as "lop leap years." Every century would thus have 75 regular years with 365 days, 24 leap years with 366 days, and a lop leap year with, again, 365 days. On average, a year would then have a length of 365.24 days.

This, however, falls short again—albeit to a small extent, but short nevertheless. A further adaptation was called for. The Pope and his advisers put their thinking caps on yet again and hit on another idea: Reinsert the leap day into every fourth lop leap year. Thus the loop would be closed, so to speak, and those years that are divisible by 400 were to be "loop lop leap years." Since the year 1600 was just around the corner, it was declared the first "loop lop leap year." The next one would be the year 2000.

Thus the average length of a year was now precisely 365.2425 days (three centuries with an average length of 365.24 days; one century with an average length of 365.25 days). Well, wouldn't you know it? This is just a wee bit too long. But by then Pope Gregory XIII had had enough. There would be no further corrections or adjustments. Not even the Church, known for its long-term planning, was prepared to go the extra . . . well, millimeter, considering the order of magnitude here discussed. The discrepancy of 26 seconds per year amounts to no more than a day every 3,322 years.

Future inaccuracies of the calendar had thus been taken care of, but what about the inaccuracies that had accumulated during the millennium and a half since Julius Caesar introduced his calendar? Pope Gregory's ingenuity solved the problem with one ingenious stroke: In 1582 he struck 10 entire days from the calendar. This drastic step had an additional benefit to the pontiff. It was an opportunity to show the world and its rulers who the true master was. So what happened was that Thursday, October 4, 1582, was followed in most Catholic countries by Friday, October 15.

But the non-Catholic countries had absolutely no intention of obeying the Pope's diktat. In England and its colonies (including America), the correction took place as late as 1752, at which point 11 days had to be struck from the calendar. The Russians stuck to their calendar until the Revolution but were then forced to cross out 13 days. The abstruse result was that the October Revolution in fact took place in November 1917.

Not all is well, however, and no one yet knows how it will all end. Even though everything seemed to run rather smoothly since Pope Gregory's adjustments, 400 years later all is threatening to fall apart again. Science has made huge progress, and atomic clocks can nowadays measure time with a precision of 10^{-14}.[1] This corresponds to an error of not more than one second every 3 million years. With such precise measurements, an annual surplus of 26 seconds becomes once again unbearable. So, what this writer wishes to propose is one further adjustment: Delete the reinserted additional leap day every eighth loop lop leap year. Hence, every 3,200 years, the month of February would again comprise only 28 days. This would be the final lap in the rounds of adjustments; hence, that particular year would be called the "lap loop lop leap year." The average year would last for 365.242188 days. Based on painstaking calculations, it turns out that the first lap loop lop leap year would be upon us in AD 4400. So we still have some time to think this whole thing over. The average year would still be one second short, but it would take 86,400 years for this error to accumulate to a full day. This is an imprecision with which even the loopiest mathematicians and church people could lope . . . I mean cope.

[1]This is a scientific way of writing 0.00000000000001.

2

Is the World Coming to an End Soon?

Isaac Newton was, as we all know, the most outstanding scientist and mathematician of the 17th and 18th centuries. He is considered the father of physics and the founder of the law of gravity. But was he really the rational thinker as we would have it today? Far from it. As it turns out, Newton was also a religious fundamentalist who devoted himself to intense Bible study and who wrote over a million words on biblical subjects.

Newton's aim was to unravel nothing less than God's secret messages. According to the great scientist, they were hidden in the Holy Scriptures. Above all Newton was intent on finding out when the world would come to an end. Then, he believed, Christ would return and set up a 1,000-year Kingdom of God on earth and he—Isaac Newton, that is—would rule the world as one among the saints. For half a century, Newton covered thousands of pages with religious musings and calculations.

Three hundred years later, toward the end of 2002, the Canadian historian of science Stephen Snobelen of King's College in Halifax found a significant document among a convoluted mass of manuscripts that had been left in the home of the Duke of Portsmouth for over 200 years. They had been kept from public scrutiny until 1936, when they were sold at an auction at Sotheby's. The collection was acquired by the Jewish scholar and collector Abraham Yehuda, an Iraqi professor of Semitic languages who, upon his death, left them to the State of Israel's Jewish National Library. Ever since, they have been gathering dust in the archives at the Hebrew University in Jerusalem.

When Snobelen assessed the manuscripts, he chanced upon a piece of paper on which the famous physicist had calculated the year of the apocalypse: 2060. Newton arrived at this date based on razor-sharp conclusions. From his readings of the Book of Daniel (Chapter 7, Verse 25) and the Book of Revelations, the physicist concluded that the time span of three and a half years signified a critical time period. Basing a year on 360 days—a simplification that comes easy to a mathematician—this time span corresponds to 1,260 days. By simply replacing days by years, the illustrious Bible researcher easily concluded that the world would come to an end 1,260 years after a particular commencing date.

So now the critical question was, what was the commencing date? Newton had several dates to choose from, each of which was somehow connected to Catholicism, a religion he loathed to the core. Richard Westfall, author of the definitive biography on Newton, pointed out that Newton had singled out the year 607 as a significant date. It referred to the year in which Emperor Phokas bestowed on to Bonifatius III the title of "Pope Over All Christians." This edict elevated Rome to *"caput omnium ecclesarum"* (head of all churches)—truly a worthy occasion to mark the beginning of the end. Since 607 + 1,260 = 1,867, Newton predicted that the end of the world would occur in the year 1867. But today we know with absolute certainty that the world did not come to an end then.

Newton had prepared for such a problem with a fallback strategy. During his research in Jerusalem, the Canadian professor also came across the year 800. This is a significant date in history since on Christmas day of that year Pope Leo III crowned Charlemagne at St. Peter's in Rome. It was the beginning of the "Holy Roman Empire of the German Nation." And 800 plus 1,260 equals 2,060. Another half century or so, and the world as we know it will call it quits. *Quod erat demonstrandum.*

If one or the other reader has started to feel slightly queasy after reading the last few lines, let him or her heave a sigh of relief. Newton had another fallback position. According to further calculations of the eminent physicist, the end of the world could be somewhat delayed and take place at the latest in the year 2370.

3

Cozy Zurich

Professors in Zurich are probably not quite aware of how lucky they are. A visiting professor, invited to give a lecture series at the University of Zurich, can attest to the fact that the circumstances that prevail in this city can be described as nothing less than paradisiacal.

As soon as one sets foot in the lecture hall, it seems that the golden age has dawned. Spanking-clean blackboards gleam in pleasant anticipation, the appropriate container is filled with virginal chalks, a clean sponge above the water basin (note: with cold and hot water) awaits its deployment, and a specially designed fastener holds a windshield wiper kind of contraption for drying the blackboard. Next to it a freshly washed and ironed kerchief dangles from a hook. It is to be used after the blackboard has been given the workover with sponge and windshield wiper in order to return to it the luster it deserves. Two overhead projectors stand ready with separate sets of scrupulously aligned, colored, felt-tipped pens next to them.

Surrounded by all this luxury, the lecturer cannot but recall with melancholy the conditions at his faraway home university. There one secures sundry materials before each lecture oneself and picks up some toilet paper on the way, in order to wipe the blackboard, if not clean, then at least partially clean when it gets too crowded. If a projector is required, the lecturer reports to the dean's office. With luck, one of the monsters is available and, after having acknowledged receipt on a special form, one is permitted to lug the contraption through interminable corridors while the extension cord keeps wrapping itself around one's legs. After the lecture, the monster, which at this point seems to have gained considerable weight, is shlepped back to the dean's office.

In Zurich, if the professor requires a computer with

special software in order to demonstrate the workings of a simulation in real time, the class need not be rescheduled to a computer lab. Rather, a friendly "house technician," dressed in dapper overalls, rolls a computer—with the necessary software having been installed the previous night—into the lecture hall at precisely the right moment, attaches it to the overhead projector, and hands the remote control to the lecturer. Mouse and keyboard wait for their cue.

Seemingly insurmountable obstacles are overcome with ease in Zurich. One hour before the planned projection of a video film, it turns out that the tape was recorded in the National Television Systems Committee (NTSC) system, which is not current in Europe. The desperate lecturer sprints down to the house technicians office where one of those gnomes of Zurich patiently explains that, first of all, there are two versions of the NTSC system; second, that projectors are of course available for the one as for the other; and, third, that *both* machines will be set up in the lecture hall, just to be on the safe side.

Before the screening of the movie, the house technician gives the lecturer a crash course in the operation of the instrument panel that is embedded in the wall next to the blackboard and which, to the uninitiated, carries an astounding similarity to the cockpit of a Boeing 747. All lights, dimmers, and switches for projectors, video machines, and computers may be controlled from this strategic command post. If, despite all precautions, something should not work to the complete satisfaction of all present, a quick call to the house technicians nerve center suffices (telephones are available on every floor, in every corridor). Within minutes a competent and friendly gentleman is on the spot to straighten everything out.

Seating arrangements in the lecture halls can be changed at request, of course. If a sociologist wants to demonstrate group dynamics, chairs and tables are moved close together. But in the break before the next lecture, a house technician swiftly aligns the furniture again, and at the sound of the bell all chairs stand at attention at their appropriate places.

It goes without saying that all students are at the ready at the start of each class. And if a blushing student, mumbling an apology, does happen to sneak in late, one is again painfully reminded of the realities at home. There the latecomers—about a third of the registered auditors in an average class (another third do not show up at all)—walk proudly into class at any time, dispense friendly greetings to the left and right before taking their seats, and do not hesitate to consult today's newspaper for important news if the content of the lecture is, in their opinion, too boring.

4

Daniel Bernoulli and His Difficult Family

More than two centuries have passed since the death of one of the most distinguished mathematicians in history: Daniel Bernoulli (1700–1782). The name Bernoulli calls for precision since the family from the Swiss town of Basle produced no fewer than eight outstanding mathematicians within three generations. Because the same given names kept being used by the family over and over again, a numbering system was adopted to tell fathers, brothers, sons, and cousins apart. It starts with Jakob I and his brother Johann I (the third brother, Nikolaus, being an artist, wasn't given a number). Then there are, in the next generation, Nikolaus I and Johann's three sons—Nikolaus II, Daniel, and Johann II. Finally, the two sons of Johann II, called Johann III and Jakob II, followed in the footsteps of their brilliant ancestors. (Johann's third son, Daniel, didn't get any further than being deputy professor at the University of Basle. Hence he wasn't given a number, which is why his famous uncle and namesake didn't need one either.)

Together with Isaac Newton, Gottfried Wilhelm Leibniz, Leonhard Euler, and Joseph-Louis Lagrange, the Bernoulli family dominated mathematics and physics in the 17th and 18th centuries. The family members were interested in differential calculus, geometry, mechanics, ballistics, thermodynamics, hydrodynamics, optics, elasticity, magnetism, astronomy, and probability theory. For more than 30 years, the Swiss National Fund has been supporting work on an edition of the complete works of Jakob I, Johann I, and Daniel. The complete edition will comprise 24 volumes. Another 15 volumes, including a selection of their 8,000 letters, is to follow.

Unfortunately, the gentlemen from Basle were as conceited and arrogant as they were brilliant and engaged constantly in rivalry, jealousy, and public rows. Actually, everything had started so idyllically. Jakob I, who had acquired his knowledge in the natural sciences as an autodidact and went on to teach experimental physics at the University of Basle, secretly introduced his younger brother to the mysteries of mathematics. This was very much against the will of their parents, who wanted the younger brother to embark on a career in commerce after the elder son had already refused to embark on the clerical career they had planned for him.

But the harmony between the two highly gifted brothers soon turned into a bitter argument. The conflict began when Jakob I, annoyed by Johann's bragging, claimed in public that the works of his former student were but copies of his own results. Next Jakob I—who by then held the chair of mathematics at the University of Basle—plotted successfully against his brother's appointment to his department. So Johann I had to teach at the University of Groningen before finally being offered a chair in Basle for . . . Ancient Greek. But fate decided otherwise, and just when Johann set off for his native town, the news reached him that Jakob had died. Thus the not-too-grief-stricken brother was given the chair of mathematics in Basle after all. Jakob's most important opus, his *Ars Conjectandi* (The Art of Conjecture), which appeared after his death, formed the basis of probability theory.

But don't believe that Johann I learned anything from this sad story. In educating his own sons he committed exactly the same mistakes as his father had done before him. Claiming that mathematics couldn't provide a living, Johann tried to bully the most gifted of his three sons, Daniel, into a career in commerce. When this attempt proved unsuccessful, he allowed him to study medicine—all this in order to prevent his son from becoming a competitor. But the sons followed the example of their elders, and Daniel, while studying medicine, took lessons in mathematics from his older brother, Nikolaus II. In 1720 he traveled to Venice to work as a physician. However, his heart belonged to physics and mathematics, and

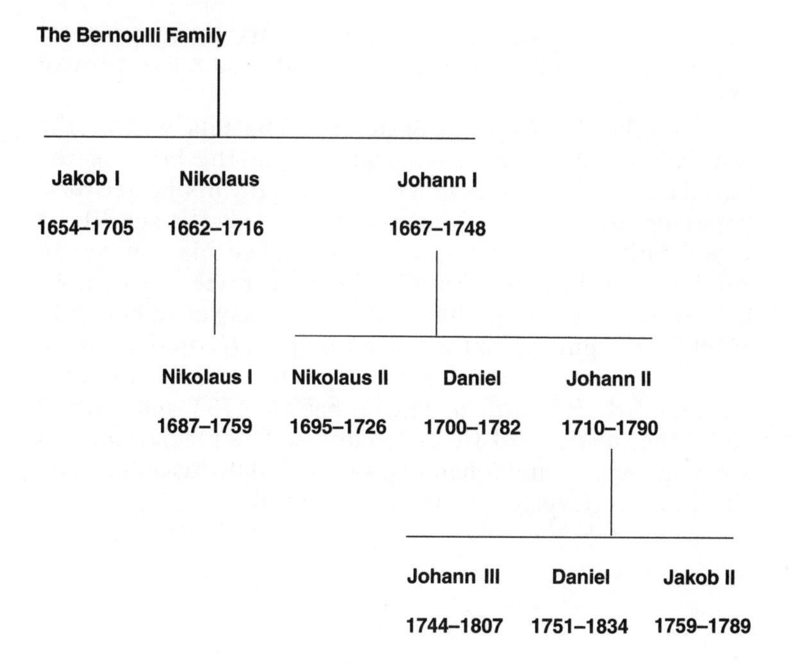

during his stay he acquired such a great reputation in these fields that Peter the Great offered him a chair at the Academy of Science in St. Petersburg.

In 1725 Daniel traveled to the capital of the Russian empire, together with his brother Nikolaus II, who had also been offered a professorship for mathematics at the academy. Their joint sojourn didn't last long. Hardly eight months after their arrival Nikolaus II fell ill with a fever and died. Daniel, who possessed more of a sense of family than his father, was very distressed and wanted to return to Basle. But Johann I didn't want to have his son back home. Instead, he sent one of his pupils to St. Petersburg to keep Daniel company. This was an extremely lucky coincidence, since the pupil was none other than Leonhard Euler, the only contemporary who could compete with the Bernoullis when it came to mathematical talent. A close friendship developed between the two Swiss mathematicians in exile. The six years they spent together in St. Petersburg were the most productive time of Daniel's life.

After he returned to Basle, the quarrels within the family started anew. When Daniel won the prize of the Parisian Academy of Science jointly with his father for a paper on astronomy, Johann I didn't exactly act like a proud father. On the contrary, he kicked his son out of the house. Daniel was to win the great prize of the academy nine times altogether. But worse was yet to come: In 1738 Daniel published his magnum opus, *Hydrodynamica*. Johann I read the book, hurriedly wrote one of his own with the title *Hydraulica*, dated it back to 1732, and claimed to be the inventor of fluid dynamics. The plagiarism was soon uncovered, and Johann was ridiculed by his colleagues. But his son never recovered from the blow.

II

Unsolved Conjectures

5

The Mathematicians' Million Dollar Baby

Henri Poincaré (1854–1912) was one of the most eminent French mathematicians of the past two centuries. Together with his German contemporary, David Hilbert, Poincaré was one of the last mathematicians who not only had a deep understanding of all areas of mathematics but also was active in all of them. After Hilbert and Poincaré, mathematics became so vast that nobody could hope to grasp more than a minute part of it.

One of Poincaré's best-known problems is what is today called the Poincaré conjecture. It has baffled and challenged several generations of mathematicians. In the spring of 2002 Michael Dunwoody from the University of Southampton believed—albeit only for a few weeks—that he had been successful in finding a proof for the conjecture.

The Poincaré conjecture is considered so important that the Clay Mathematics Institute named it one of the seven Millennium Prize problems. Each person who is the first to solve one of the problems will be awarded $1 million. The prize committee believed it would take decades for the first of the prizes to be allocated; but here, a mere two years after the announcement, the Clay Foundation seemed to be faced with the possibility of having to award its first prize. But doubts arose about the validity of the proof provided by Dunwoody—and as it turned out there was ample reason.

The Poincaré conjecture falls within the realm of topology. This branch of mathematics focuses, roughly speaking, on the issue of whether one body can be deformed into a different body through pulling, squashing, or rotating, without tearing or gluing pieces together. A ball, an egg, and a flowerpot are, topologically speaking, equivalent bodies, since any one of them can be deformed into any of the others without performing any of the "illegal" actions. A ball and a coffee cup, on the other hand, are not equivalent, since the cup has a handle, which could not have been formed out of the ball without poking a hole through it. The ball, egg, and flowerpot are said to be "simply connected" as opposed to the cup, a bagel, or a pretzel. Poincaré sought to investigate such issues not by geometric means but through algebra, thus becoming the originator of "algebraic topology."

In 1904 he asked whether all bodies that do not have a handle are equivalent to spheres. In two dimensions this question refers to the surfaces of eggs, coffee cups, and flowerpots and can be answered *yes*. (Surfaces like the leather skin of a football or the crust of a bagel are two-dimensional objects floating in three-dimensional space.) For three-dimensional surfaces in four-dimensional space, the answer is not quite clear. While Poincaré was inclined to believe that the answer was *yes*, he was not able to provide a proof.

Interestingly enough, within several decades mathematicians were able to prove the equivalent of Poincaré's conjecture for all bodies of dimension greater than four. This is because higher-dimensional spaces provide more elbowroom, so mathematicians find it simpler to prove the Poincaré conjecture. Christopher Zeeman in Cambridge started the race in 1961 by proving Poincaré's conjecture for five-dimensional bodies. In the same year Stephen Smale from Berkeley announced a proof for bodies of seven and all higher dimensions. John Stallings, also from Berkeley, demonstrated a year later that the conjecture was correct for six-dimensional bodies. Finally, in 1982, Michael Freedman from San Diego provided proof for four-dimensional bodies. All that was left now were three-dimensional bodies floating in four-dimensional space. This was all the more frustrating since four-dimensional space represents the space-time continuum in which we live.

Michael Dunwoody thought he had found a proof. On April 7, 2002, he posted a preprint entitled "Proof for the Poincaré Conjecture" on the Web. Reputable mathematicians called it the first serious attempt at solving the Poincaré conjecture for a long time. In higher dimensions it is really not at all easy to recognize a sphere when you meet one, despite all the additional elbowroom. To understand the difficulty, just think of olden times, when pirates and adventurers did not realize that the world was round, despite all their expeditions and discovery trips. Dunwoody based his work on previous work by Hyam Rubinstein, an Australian mathematician who had studied the surfaces of four-dimensional spheres. (Remember: The surface of a four-dimensional object is a three-dimensional object.)

Dunwoody required no more than five pages to develop his argument, which concluded with the statement that all simply connected, closed, three-dimensional surfaces can be converted into the surface of a sphere by means of pulling, squishing, and squashing—but without tearing. This statement is equivalent to a proof of Poincaré's conjecture.

Alas, only a few weeks after posting his findings on the Web, Dunwoody was forced to append a question mark to the title of his paper. One of his colleagues had discovered that the proof had a hole. The title now read "Proof for the Poincaré Conjecture?" and, though Dunwoody immediately attempted to fill the hole, he was unsuccessful. So were friends and colleagues who tried the same. A few weeks later the paper disappeared from the Web. Poincaré's conjecture remained as elusive as ever. (See, however, Chapter 13.)

6

A Puzzle by Any Other Name

One day in the mid-1980s, Jeff Lagarias, a mathematician employed by AT&T, gave a lecture about a problem on which he had spent a considerable amount of time but for which he had found no solution. In fact, he had not even come close. It was a dangerous problem he continued, obviously speaking from experience, because those who work on it were in danger of compromising their mental and physical health.

What is this dangerous problem?

In 1932 Lothar Collatz, a 20-year-old German student of mathematics, came across a conundrum that, at first glance, seemed to be nothing more than a simple calculation. Take a positive integer x. If it is even, halve it ($x/2$); if it is odd, multiply it by 3, add 1, and then halve it: ($3x + 1)/2$. Then, using the result, start over again. Stop if you hit the number 1; otherwise continue.

Collatz observed that starting from any positive integer, repeated iterations of this procedure sooner or later lead to the number 1. Take, as an example, the number 13. The resulting sequence consists of the numbers 20, 10, 5, 8, 4, 2, and 1. Take 25. You get 38, 19, 29, 44, 22, 11, 17, 26, 13, 20, 10, 5, 8, 4, 2, and 1 again. No matter which starting number Collatz tested, he always ended up with the number 1.

The young student was taken aback. The number sequence could have just as easily veered off toward infinity or entered an endless cycle (not containing 1). At least that should have happened occasionally. But *no*, the sequences ended up at 1 every single time. Collatz suspected that he might have discovered a new law in number theory. Without further ado he set about seeking a proof for his conjecture. But his efforts amounted to nought. He managed neither to prove his conjecture nor to find a counterexample, that is, a number sequence that does *not* end with 1. (In mathematics it suffices to find one counterexample to disprove a conjecture.) Throughout his life Collatz was unable to publish anything noteworthy about this conjecture.

Some time during the Second World War, the problem was picked up by Stanislaw Ulam, a Polish mathematician who held a high position with the Manhattan Project. To while away his free time (there was not a whole lot to do in Los Alamos in the evenings), Ulam investigated the conjecture but failed to find a proof. What he did do was describe it to his friends, who from then on called it Ulam's problem.

Another few years went by and Helmut Hasse, a number theorist from the University of Hamburg, stumbled over this curious puzzle. Caught by the bug, he gave lectures about it in Germany and abroad. One attentive member of the audience observed that the number sequence goes up and down like a hailstone in a cloud before it invariably plummets to earth. Time for a name change then—the number sequence was henceforth called the Hailstone sequence and the prescription to calculate it the Hasse algorithm. When Hasse mentioned the problem during one of his lectures at Syracuse University, the audience at that particular event named it the Syracuse problem.

Then the Japanese mathematician Shizuo Kakutani lectured on the topic at Yale University and the University of Chicago, and the problem immediately became known as the Kakutani problem. Kakutani's lectures set off efforts by professors, assistants, and students as feverish as they were futile. Proof eluded everyone. Thereupon a rumor started to make the rounds that the problem was, in fact, an intricate Japanese plot intended to put a brake on the progress of American mathematics.

In 1980 Collatz, whose initial contribution had been all but forgotten, reminded the public that it was he who had discovered the sequence. In a letter to a colleague he wrote: "Thank you very much for your letter and your interest in the function which I inspected some fifty years ago." He went on to explain that at the time there had only been a table calculator at his disposal, and he had therefore been unable to calculate the Hailstone sequence for larger numbers. As a postscript he added: "If it is not too immodest, I would like to mention that at the time Professor H. Hasse called the conundrum the 'Collatz problem'."

In 1985 Sir Bryan Thwaites from Milnthorpe in England published an article that left little doubt as to who he thought the author of the conjecture was. The article was entitled "My Conjecture." Thwaites went on to claim that he had been the originator of the problem three decades earlier. In a separate announcement in the *London Times*, he proposed a prize of £1,000 to whomever would be able to provide a rigorous proof for what henceforth should be called the Thwaites conjecture.

In 1990 Lothar Collatz, who had made a name for himself as a pioneer in the field of numerical mathematics, died shortly after his 80th birthday. He was never to find out whether the conjecture that—he would have been happy to know—is nowadays usually called the Collatz conjecture was true or false.

In the meantime mathematics had found a new tool—the computer. Today just about anybody can prove on his or her PC that Collatz's conjecture is correct for the first few thousand numbers. In fact, with the help of supercomputers, all numbers up to 27 quadrillion (that is, 27 followed by 15 zeros) have been tested. Not one was found whose Hailstone sequence did not end with 1.

Such numerical calculations do not represent a proof, of course. All they achieved was the discovery of several historical records, one of which was the discovery of the longest Hailstone sequence to date: a certain 15-digit number whose Hailstone sequence consists of no less than 1,820 numbers before it reaches the final position of 1. One thing Jeff Lagarias did manage to prove in the course of his frustrating endeavors was that a counterexample—should one exist—would have to have a cycle comprising at least 275,000 way points.

So a computer is of little help in finding a counterexample to the Collatz conjecture. In the final analysis, the question is not decidable by computer anyway because only numbers that fulfill Collatz's conjecture, that is, whose Hailstone sequences go to 1, will induce the computer program to stop. If indeed there exists a counterexample —either because its Hailstone sequence tends toward infinity or because it enters a very long cycle that does not contain 1—the computer program would simply produce numbers, without terminating. A mathematician sitting in front of the computer monitor would never have a way of knowing whether the sequence eventually escapes toward infinity or starts on a cycle. At some point he would probably simply hit the Escape key and go home.

7

Twins, Cousins, and Sexy Primes

A scientific lecture at the Research Institute for Mathematics in Oberwolfach, Germany, is no uncommon event. Nevertheless, the lecture by the American mathematician Dan Goldston, of the State University of San Jose, in spring 2003 was of a completely different order. Its contents took the mathematical community by storm. Together with his Turkish colleague Cem Yildirim, it looked like Goldston had advanced attempts to prove the so-called twin primes conjecture by a significant step. What is it about these quirky siblings that so excites mathematicians?

Within the group of integers, prime numbers are in a way thought of as atoms, since all integers can be expressed as a product of prime numbers (for example, 12 = 2 × 2 × 3), just as molecules are made up of separate atoms. The theory of prime numbers continues to be shrouded in mystery and still holds many secrets. Here is just one example: In 1742 Christian Goldbach and Leonhard Euler formulated the unproven Goldbach conjecture, which says that every even integer greater than 2 can be expressed as the sum of two primes (for example, 20 = 3 + 17).

While the chemical periodic table comprises only 10 dozen elements out of which all materials consist, the two ancient Greek mathematicians Euclid and Eratosthenes already knew that there was an infinite number of primes. The most important question is, how are the primes distributed within the system of integers? Taking the first 100 numbers we count 25 primes; between 1,001 and 1,100 there are only 16; and between the numbers 100,001 and 100,100 there are a mere six. Prime numbers become increasingly sparse. In other words, the average distance between two consecutive primes becomes increasingly large (hence the phrase "they are few and far between").

Around the turn of the 19th century, the Frenchman Adrien-Marie Legendre and the German Carl Friedrich Gauss studied the distribution of primes. Based on their investigations they conjectured that the space between a prime P and the next bigger prime would, on average, be as big as the natural logarithm of P.

The value obtained, however, holds true only as an average number. Sometimes the gaps are much larger, sometimes much smaller. There are even arbitrarily long intervals in which no primes occur whatsoever. The smallest gap, on the other hand, is two, since there is at least one even number between any two primes. Primes that are separated from each other by a gap of only two—for instance, 11 and 13, or 197 and 199—are called twin primes. There are also prime cousins, which are primes separated from each other by four nonprime numbers. Primes that are separated from each other by six nonprime numbers are called, you guessed it, sexy primes.

Much less is known about twin primes than about regular primes. What is certain is that they are fairly rare. Among the first million integers there are only 8,169 twin prime pairs. The largest twin primes so far discovered have over 50,000 digits. But much is unknown. Nobody knows, for instance, whether an infinite number of twin prime pairs exist, or whether after one particular twin prime pair there are no larger ones. Mathematicians believe that the first case holds true, and this is what Goldston and Yildirim set out to prove.

What they claimed was that an infinite number of gaps exist between consecutive primes that are much, much smaller than the logarithm of P, even if P tends toward infinity. The two mathematicians were not given much time to rejoice over their findings. No sooner was it announced than they were awoken to reality. Two colleagues had decided to retrace their proof step by step. In the course of painstaking work, they noticed that Goldston and Yildirim had neglected an error term, even though the term was too large. This was unacceptable and invalidated the entire proof.

Two years later, with the help of Janos Pintz from

Hungary, Goldston and Yildirim revised their work. They managed to plug the hole, and their proof is now believed to be correct. Even though their work does not prove that there are an infinite number of twin prime pairs, it is certainly a step in the right direction.

Working on the theory of twin primes may be more than just an intellectual exercise, as Thomas Nicely from Virginia discovered in the 1990s. Hunting for large twin prime pairs, he was running through all integers up to 4 quadrillion. The algorithm required the computation of the banal expression x times $(1/x)$. But to his shock, when inserting certain numbers into this formula, he received not the value 1 but an incorrect result. On October 30, 1994, Nicely sent an e-mail to colleagues to inform them that his computer consistently produced erroneous results when calculating the above equation with numbers ranging between 824,633,702,418 and 824,633,702,449. Through his research on twin primes, Nicely had hit on the notorious Pentium bug. The error in the processor cost Intel, the manufacturer, $500 million in compensations—a prime example (no pun intended) that mathematicians can never tell where their research, and errors, may lead them.

8

Hilbert's Elusive Problem Number 16

Mathematical proofs, by their nature, are often very complex, and ascertaining whether they are, in fact, correct requires painstaking efforts by experts. A case in point is what happened on March 28, 2003, when Dan Goldston and Cem Yildirim, American and Turkish mathematicians respectively, believed they had achieved a major breakthrough with the so-called twin prime conjecture. Celebration turned into disappointment only a few weeks later when, on April 23, colleagues announced that they had found a hole in their argument. A year earlier the Englishman M. J. Dunwoody had presented a proof for the Poincaré conjecture. Here, too, not more than a couple of weeks went by before the proof turned out to be incomplete. A third case in point is Andrew Wiles's proof for Fermat's theorem. During the refereeing process it was discovered that the proof was incomplete. In this instance the error was repairable, but it took one and a half years and the help of a willing colleague to plug the hole.

Old, unsolved problems, especially those associated with the names of famous mathematicians, exert immense fascination. Pondering problems that experts from previous centuries dealt with has its own appeal. The 23 problems that David Hilbert, the famous mathematician from Göttingen, listed in 1900, which were to determine the direction of mathematical research for the better part of the next century, are engulfed by that same aura of mystery. By now solutions have been found for 20 of them, but numbers 6 (the axiomatization of physics), 8 (the Riemann conjecture), and 16 have so far eluded the mathematical world.

Indeed, numbers 8 and 16 are considered of such importance that the mathematician Steve Smale listed them among the most important mathematical problems for the 21st century. But as in many cases, fascination is closely associated with danger. Famous problems work their magic also on people who do not possess the prerequisites to deal with them in a proper fashion. And, like falling in love with the wrong person, once one is hooked, the danger of self-deception regarding one's own suitability is great.

Caution was therefore indicated when the news broke in November 2003 that a 22-year-old female student, Elin Oxenhielm from Sweden, had cracked part of Hilbert's 16th problem. Her work had been checked by referees from the mathematics journal *Nonlinear Analysis* and accepted for publication. Elin Oxenhielm, immensely proud that her first piece of work was an absolute hit, immediately notified the media. Even though her department had advised her to let caution prevail, she proceeded to give interviews, announced plans for a book, and did not even rule out a film about Hilbert's 16th problem. A brilliant career seemed to be in the bag, positions at leading institutes were in the offing, and a steady stream of funding was certain.

Hilbert's 16th problem deals with two-dimensional dynamical systems. The solutions of such systems can reduce to single points or end in cycles. Hilbert investigated differential equations that describe such dynamical systems, equations whose right-hand sides were made up of polynomials.[1] He asked how the number of cycles depended on the degree of the polynomials. An answer would be of particular interest for complex or chaotic systems.

Oxenhielm's eight-page paper begins with the observation that in simulations a certain differential equation behaves like the trigonometric sine function. Then she approximates the equation, without an estimation of even

[1]A polynomial is an expression like $x^4 + 5x^3 + 7x^2 + 2x$. The degree of this polynomial is 4.

the order of magnitude of the neglected terms. After recasting the equation a few times, a further approximation is made and justified by no more than numerical examples and computer simulations. Finally, in an unproven assertion, Oxenhielm claims that the results are not falsified by the approximations. Such nonchalant use of the mathematical rules of the game renders her work completely and utterly useless.

The media reports that Oxenhielm had brought about not only informed the public of her feat but also alerted professional circles. One incensed expert wrote an outraged letter to *Nonlinear Analysis* with the urgent request to stop the intended publication. Elin Oxenhielm's supervisor at the university, who had previously read and criticized her findings, asked the editors to delete her name from Oxenhielm's list of acknowledgments. She did not want to be associated in any way with the paper. To add insult to injury, a technical university gave its freshman students a homework assignment to list the shortcomings in Oxenhielm's paper.

The avalanche of criticism had its effects. On December 4, 2003, the publishers of *Nonlinear Analysis* announced they were postponing publication of Oxenhielm's work pending further review. Shortly thereafter, the paper was withdrawn from publication.

How could things get that far? It is certainly not unusual for an inexperienced scientist to submit a paper that has errors or gaps to a journal. Usually the refereeing process of a journal ensures that mistakes come to light and inferior work does not get published. This is why journals with a good reputation often reject 90 percent or more of submitted papers. But in this particular case the process broke down completely. One expert who was interviewed believed that the journal's referees, whose identities usually remain confidential, may have been engineers for whom approximations are common practice, so long as they do not cause problems. But to proceed in such a manner in the field of mathematics is unacceptable.

Furthermore, the fact that the young woman approached the media was quite unforgivable. It is the unfortunate destiny of mathematicians that for most of their lives they sit all by themselves in little rooms, solving centuries-old problems. Only on rare occasions is the public notified of a success. Working under such circumstances, secluded from the hustle and bustle of the outside world, ensures a certain level of quality. Since mathematical proofs can only be considered correct after they have withstood the test of time, media fanfare is detrimental to the painstaking and protracted examination of a proof. Actively inviting this kind of public relations is unbecoming, to say the least.

Often the only satisfaction a mathematician may expect from a successful proof is acknowledgment from colleagues in the same field. Experts may number no more than a dozen, spread over the farthest corners of the globe. Receiving their e-mailed nods of approval often represents the height of approbation. On the rare occasions where solid applications for a mathematical theorem appear, they become public knowledge only many decades later. That young researchers, frustrated by the prospect of a life in the shadows of anonymity, seek out the public arena may be understandable. Nevertheless, most mathematicians avoid the limelight. To inform the media at the drop of a hat conveys an image from which the protagonists shy away. Mathematicians' subtle trains of thought, minute considerations, and rigorous arguments do not lend themselves to infotainment. For better or worse, mathematics is a science with a low profile.

Solved Problems

9

The Tile Layer's Efficiency Problem

Every month around 4,000 articles, written by researchers around the world, are published in countless scientific journals. In January 2002 the American Mathematical Association (AMA) chose to highlight a paper written by the American mathematician Thomas Hales for its significance.

For once a strictly theoretical mathematical piece could possibly also appeal to tradesmen. Tile layers who cover bathroom, kitchen, and porch floors with tiles of all shapes and forms may find the paper of interest. Maybe one or the other among them has pondered the question as to which shape has the smallest perimeter while covering the same region as other tiles. This is the very issue addressed in the article selected as outstanding by the AMA for its mathematical, if not its practical, relevancy.

A tile layer can take tiles of triangular, square, pentagonal, hexagonal, heptagonal, and octagonal shape—all having the same surface area—measure their perimeters and verify which has the smallest circumference. So far, so good. But it would be too soon to start mixing the grout. If the tile layer tried to cover the kitchen floor with pentagonal tiles, he would soon notice that gaps appear between the tiles. Pentagons are unusable as floor coverings because, when aligned next to each other, they do not fit together seamlessly. It is the same story for heptagons, octagons, and most other regular polygons—in none of these cases can the kitchen floor be completely covered without leaving any spaces in between the tiles.

The ancient Pythagoreans were quite familiar with this fact of geometry. They knew that among all regular polygons, only triangles, squares, and hexagons could cover an area without leaving any spaces between them. Any other regular polygon invariably produces gaps.

So the tile layer's choices are quite limited. All he can do is examine which of the three admissible shapes has the smallest perimeter. Take an area of 100 square centimeters: The triangular-shaped tiles have a circumference of 45 centimeters, the squares a circumference of 40 centimeters, and the hexagons—with a mere 37 centimeters—have the smallest circumference. Pappus of Alexandria (approximately AD 290–350) was already aware that hexagons were the most efficient regular polygons. So were honeybees. They want to store as much honey as possible in containers using the least possible amount of wax. So they build honeycombs in hexagonal shapes.

The reason the hexagon has the smallest perimeter is that, of the three possible tile shapes, it is most similar to the circle. And among all geometric shapes the circle has the smallest perimeter. To encircle an area of 100 square centimeters, the circle requires a perimeter of only about 35 centimeters.

Can we now claim that the problem is solved? By no means. Who says that the floor covering must consist of a single tile shape? And why should the shape have to be regular or straight edged? Indeed, the tiles do not even have to be convex; imagine tiles that bulge inward or outward. So floors could very well be tiled with varying shapes that would, by the way, only enhance their aesthetic appeal, as M. C. Escher has so masterfully shown us in his prints.

The general question that mathematicians have asked themselves is, which tile taken from the multitude of imaginable forms and shapes has the smallest perimeter? For 1,700 years it has been conjectured that the solution is the hexagonal honeycomb. All that was missing was a proof.

Hugo Steinhaus (1887–1972), a Polish mathematician from Galicia, was the first to make significant inroads. He proved that as far as tiles that consisted of only a single shape were concerned, hexagons represented the least-perimeter way to cover a floor. This was progress since, in contrast to Pappus, Steinhaus also allowed irregular-shaped tiles. In 1943 the Hungarian mathematician László Fejes Toth (1915–2005) took the next step. He proved that among all convex polygons it was the hexagon that had the smallest perimeter. Contrary to Steinhaus, Fejes Toth did not require the floor to be covered with just one kind of tile but permitted the use of collections of differently shaped tiles. And yet his theorem ignored tiles that were not straightedged.

A completely general proof was provided only in 1998 by Thomas Hales. Just a few weeks earlier he had solved the oldest open problem in discrete geometry, the 400-year-old Kepler's conjecture. The question was how identical spheres could be packed as tightly as possible. Hales proved that the densest way to pack spheres is to stack them in the same manner as grocers stack oranges: Arrange them in layers, with each sphere resting in a small hollow between three spheres beneath it. Hale's proof made headlines throughout the world. But the young professor did not waste his time bathing in glory.

On August 10, 1998, the Irish physicist Denis Weaire of Trinity College in Dublin read the news in the newspaper. Without wasting any time, he sent Hales an e-mail in which he drew his attention to the honeycomb problem and added the challenge: "It seems worth a try."

Fascinated, Hales set to work. To prove Kepler's conjecture, he had spent five years and literally burned the fuses of computers. In comparison, the new problem was a picnic. He merely required pencil and paper and half a year's work.

Hales started by dividing the infinitely large floor space into configurations of finite size. Then he developed a formula that brought a tile's area into relation to its perimeter. Next, he turned his attention to convex shapes. For every convex tile—a tile that bulged outward—there had to be corresponding tiles that bulged inward. With the help of the area-to-perimeter formula, Hales was able to prove that tiles that bulged inward require more perimeter than was saved by outward-bulging tiles. Overall this meant that round-edged polyhedra provide only disadvantages. They were thus ruled out as contenders for the title of smallest-perimeter tiles.

Since only straightedged tiles remained as candidates, the rest was clear. After all, Fejes Toth had already proven that regular hexagons represent the best combination of tiles from among all straightedged polygons. Thus Hales had provided conclusive proof that bees do exactly the right thing when constructing hexagonally shaped honeycombs.

10

The Catalanian Rabbi's Problem

Problems in number theory can usually be stated quite simply. Even a toddler might know that 9 minus 8 equals 1. Most primary school children know that $9 = 3 \times 3$ and that $8 = 2 \times 2 \times 2$. Finally, most secondary school children know that 9 equals 3^2 and 8 equals 2^3. This allows us to see another way of writing the equation $9 - 8 = 1$—namely, $3^2 - 2^3 = 1$. Is it possible to formulate probing questions about such a simple and innocent equation? As it turns out, the answer is an unequivocal *yes*. As unbelievable as it may sound, this innocent-looking equation has been the source of puzzlement for one and a half centuries.

In 1844 the mathematics periodical *Crelle's Journal* published a query by the Belgian mathematician Eugène Charles Catalan. It asked whether apart from the numbers 2 and 3 there exist integers x, y, u, and v, all greater than 1, that provide a solution to the equation $x^u - y^v = 1$ (just as 2 and 3 do in the equation $3^2 - 2^3 = 1$). Catalan suggested that there were none but could not provide proof for this.

Deceptively simple as this conjecture might seem, its solution is all the more complex. That u and v would have to be primary numbers was realized fairly soon, but after that there was no progress on the question for 158 years. Only in the spring of 2002 did something happen. The mathematician Preda Mihailescu, of the University of Paderborn in Germany, found the key to this conjecture.

How did he do it? For the Rumanian-born mathematician it all began in Switzerland, at Zurich's venerable *Eidgenössische Technische Hochschule*. This renowned institution was where Mihailescu acquired the necessary mathematical tools for his later discoveries. But just before he put the finishing touches to his doctoral thesis, he decided to switch from university to industry. He would return to academia only later, at which point he decided to embark on a second doctoral thesis. The topic was prime numbers, and this one Mihailescu actually finished. But it was while working for a high-tech firm as an expert on fingerprints that Mihailescu encountered the so-called Catalan conjecture for the first time.

In the early 14th century, more than 500 years before Catalan formulated the conjecture in *Crelle's Journal*, Levi Ben Gerson, a Jewish scholar known as Leo Hebraeus—who lived in Catalonia of all places—mentioned a variant of this problem. The rabbi proved that 8 and 9 are the only powers of 2 and 3 that differ by 1. Four centuries later Leonhad Euler demonstrated that the conjecture was true if the powers—u and v in the formula—were limited to the integers 2 and 3. And then it all went quiet again until 1976, the year in which the next step toward progress was made.

Based on the seminal work of the Cambridge mathematician Alan Baker, the Dutchman Robert Tijdeman, of the University of Leiden in the Netherlands, was able to prove that there could only be a finite number of solutions, should any exist at all, to the above equation. In the very same year it was demonstrated that in this case the powers would have to be smaller than 10^{110}.

Even though this is an astronomically large number—a 1 followed by 110 zeros—the result opened the floodgates. From then on it was just a question of lowering this upper limit on the potential solutions to a manageable number and then testing Catalan's equation running through all exponents. Maurice Mignotte, of the University Louis Pasteur in Strasbourg, France, was the first to lower the bar. In 1999 he demonstrated that the exponents of potential solutions had to be less than 10^{16}. At the same time it was already proven that they had to be larger than 10^7. The range had thus been significantly reduced but was still far too large for a computer-assisted solution to the problem.

It was time for Mihailescu's first strike. While daydreaming on a boring train ride to Zurich after a conference in Paris, an idea suddenly hit him: The exponents in Catalan's equation had to be Wieferich pairs, two numbers that can be divided by each other in a somewhat complicated manner. Wieferich pairs are very rare, and so far only six of them have been found. Thus the hunt for possible solutions to Catalan's equation could be limited to Wieferich pairs that, furthermore, had to be smaller than 10^{16}. With one stroke the problem had become amenable to computer verification. A project was launched whereby Internet users could put the idle times of their PCs to work, hunting for Wieferich pairs and testing them on Catalan's equation. But the search progressed very slowly, and in 2001 the project was aborted. By this time the lower limit had at least been raised to 10^8, but testing even this reduced range of numbers—10^8 to 10^{16}—would take years.

Now was the time for Mihailescu's second strike. He recalled an obscure subject called the "theory of cyclotomic fields" that had been developed by the German mathematician Eduard Kummer (1810–1893) in a futile attempt to prove Fermat's conjecture. Thus, a century later, Mihailescu was able to draw on Kummer's spadework to plug the last hole in the proof of Catalan's conjecture.

How does one feel after solving an age-old, world-renowned problem? It was no real high, Mihailescu recounts. After having been convinced on half a dozen earlier occasions that he had achieved his goal, only to find a gaping hole shortly thereafter, he had become very cautious. With time he just slowly slithered into the certitude that he had found success at last. So he showed the proof to his colleague Mignotte, who had spent half a lifetime on the problem. The next morning Mignotte informed him that he thought the proof was correct. They did not rejoice, but they were very happy.

11

Even Infinite Series End Sometimes

The sum of the infinite number sequence 1, 1/2, 1/4, 1/8, . . . tends toward the value 2, something that can easily be guessed by adding the first few entries of the series. But this should not lead you to believe that every infinite number sequence whose entries decrease in value has a finite sum. The so-called harmonic series, for example, which starts as 1 + 1/2 + 1/3 + 1/4 + 1/5 + . . . tends toward infinity. It does so very slowly: No less than 178 million entries must be added to arrive at the sum of 20. In mathematical jargon one says that the harmonic series diverges. Infinite number sequences whose entries add up to a finite sum are called convergent.

During the Enlightenment, number sequences and their sums were considered important areas of research. In 1644 the 19-year-old student Pietro Mengoli from Bologna, who later became a priest and professor of mathematics, asked whether the sum of the sequence consisting of reciprocal squared numbers (1, 1/4, 1/9, 1/16 . . .) converges and, if so, toward which value.

Over the years Mengoli amassed vast experience working with infinite number sequences. It was he, for example, who proved that the harmonic series diverged but that the alternating harmonic series—where the entries are added and subtracted on an alternating basis—converges toward 0.6931. But he had no answer for the series of reciprocal squared numbers. He had the suspicion that the sum approaches the approximate value 1.64, but even that he was not entirely sure about.

Some years later the mathematician Jakob Bernoulli, from the Swiss town of Basle, caught wind of this mysterious number sequence. The scientist, famous throughout Europe for his mathematical abilities, could not find a solution either. Frustrated, he penned a notice in 1689 in which he wrote: "If anybody finds out anything and is good enough to inform us, we would be very grateful."

At the turn of the 18th century, European intellectuals were absolutely fascinated by this particular problem. The number sequence and the mystery surrounding it became chosen topics of conversation in the salons of the social elite. Soon it was considered as significant as the Fermat problem, which by then was already 50 years old. Several mathematicians, among them the Scotsman James Stirling, the Frenchman Abraham de Moivre, and the German Gottfried Wilhelm Leibniz, cut their teeth on it. In 1726 the problem returned to Switzerland, to its hometown Basle.

Jakob Bernoulli's brother Johann, who was a well-known mathematician in his own right, had an exceptionally gifted student named Leonhard Euler, also from Basle. Euler was regarded as a rising star in the world of mathematics. To encourage him, Johann asked him to work on the problem. Due to its connection with mathematicians from Basle, the problem of reciprocal squared numbers became known as the Basle problem.

Euler spent many years working on the problem, sometimes putting it aside for months only to pick it up again. Finally, in the autumn of 1735 he believed he had found the solution. Nearly half a century after Pietro Mengoli had first thought about this number sequence, Euler stated that the value of the sum, calculated to the sixth digit, was 1.644934.

What led him to this solution? Surely he did not just add up the entries of the sequence. To calculate the sum to only five digits, Euler would have had to consider over 65,000 numbers. Obviously, the Swiss mathematician had guessed the exact value of the sum—which turns out to be π squared, divided by 6—even before he was able to prove it.[1] For a while Euler refused to announce the solution to the public, since even he was surprised by the

[1] π is the greek letter pi, pronounced "pie."

result. What on earth does π, the proportion of the circle's circumference to its diameter, have in common with the sum?

With the publication of *De Summis Serierum Reciprocarum* a few weeks later, Euler provided the proof to his assertion. In it he wrote that he had "quite unexpectedly found an elegant formula for $1 + 1/4 + 1/9 + 1/16$ which is dependent on the squaring of the circle!" Johann Bernoulli was at once startled and relieved. "Finally my brother's burning desire has been fulfilled," he said. "He had remarked that investigating the sum was much more complex than anyone would have thought. And he had openly admitted that all his efforts had been in vain."

The solution was unexpected because Euler had chanced on it while studying trigonometric functions. The so-called series expansion of the sine function is closely related to the reciprocal squared number series. And since trigonometric functions are related to circles, the number π occurs as part of the solution.

Euler's proof established a relationship between number series and integral calculus, which at the time was still a young branch of mathematics. Today, it is well known that the Basle number series represents a special case of a more general function (the zeta function), which plays a significant part in modern mathematics.

12

Proving the Proof

In August 1998, Thomas Hales sent an e-mail to dozens of mathematicians in which he declared that he had used computers to prove a conjecture that had evaded certain confirmation for 400 years.[1] The subject of his message was Kepler's conjecture, proposed by the German astronomer Johannes Kepler, which states that the densest arrangement of spheres is one in which they are stacked in a pyramid—much the same way grocers arrange oranges. Soon after Hales made his announcement, reports of the breakthrough appeared on the front pages of newspapers around the world. But Hales's proof remains in limbo. He submitted it to the prestigious *Annals of Mathematics*, but it is yet to appear in print. Those charged with checking it say that, while they believe the proof is correct, they are so exhausted with the verification process they cannot definitively rule out any errors. So when Hales's manuscript finally does appear in the *Annals*, it will carry an unusual editorial note—a statement that parts of the paper have proved impossible to check.

At the heart of this bizarre tale is the use of computers in mathematics, an issue that has split the field. Sometimes described as a "brute force" approach, computer-aided proofs often involve calculating thousands of possible outcomes to a problem in order to produce the final solution. Many mathematicians dislike this method, arguing that it is inelegant. Others criticize it for not offering any insight into the problem under consideration. In 1977, for example, a computer-aided proof for the four-color theorem, which states that no more than four colors are needed to fill in a map so that any two adjacent regions have different colors, was published. No errors have been found

[1]This is the same Thomas Hales we encountered in Chapter 9.

in the proof, but some mathematicians continue to seek a solution using conventional methods.

Hales (who worked on his proof at the University of Michigan in Ann Arbor before moving to the University of Pittsburgh, in Pennsylvania) began by reducing the infinite number of possible stacking arrangements to 5,000 contenders. He then used computers to calculate the density of each arrangement, which was more difficult than it sounds. The proof involved checking a series of mathematical inequalities using specially written computer code. In all more than 100,000 inequalities were verified over a 10-year period. Robert MacPherson, a mathematician at the Institute for Advanced Study in Princeton, New Jersey, and an editor of the *Annals,* was intrigued when he heard about the proof. He wanted to ask Hales and his graduate student Sam Ferguson, who had assisted with the proof, to submit their finding for publication, but he was also uneasy about the computer-based nature of the work.

The *Annals* had, however, already accepted a shorter computer-aided proof, a paper on a problem in topology. After sounding out his colleagues on the journal's editorial board, MacPherson asked Hales to submit his paper. Unusually, MacPherson assigned a dozen mathematicians to referee the proof—most journals tend to employ one to three. The effort was led by Gábor Fejes Toth, of the Alfréd Rényi Institute of Mathematics in Budapest, Hungary, whose father, the mathematician László Fejes Toth, had predicted in 1965 that computers would one day make a proof of Kepler's conjecture possible. It was not enough for the referees to rerun Hales's code; they had to check whether the programs did the job they were supposed to do. Inspecting all of the code and its inputs and outputs, which together take up 3 gigabytes of memory space, would have been impossible. So the referees limited themselves to consistency checks, a reconstruction of the thought processes behind each step of the proof, and then a study of all of the assumptions and logic used to design the code. A series of seminars, which ran for whole academic years, were organized to aid the effort.

But success remained elusive. In July 2002, Fejes Tóth reported that he and the other referees were 99 percent certain that the proof was sound. They found no errors or omissions but felt that, without checking every line of code, they could not be absolutely certain the proof was correct.

For a mathematical proof, this was not enough. After all, most mathematicians believe in the conjecture already; the proof is supposed to turn that belief into certainty. The history of Kepler's conjecture also gives reason for caution. In 1993, Wu-Yi Hsiang, then at the University of California, Berkeley, published a 100-page proof of the conjecture in the *International Journal of Mathematics*. But shortly after publication, errors were found in parts of the proof. Although Hsiang stands by his paper, most mathematicians do not believe it is valid.

After the referees' reports had been considered, Hales says that he received the following letter from MacPherson: "The news from the referees is bad, from my perspective. They have not been able to certify the correctness of the proof, and will not be able to certify it in the future, because they have run out of energy. . . . One can speculate whether their process would have converged to a definitive answer had they had a more clear manuscript from the beginning, but this does not matter now."

The last sentence lets some irritation shine through. The proof that Hales delivered was by no means a polished piece. The 250-page manuscript consisted of five separate papers, each a sort of lab report that Hales and Ferguson filled out whenever the computer finished part of the proof. This unusual format made for difficult reading. To make matters worse, the notation and definitions also varied slightly between the papers.

MacPherson had asked the authors to edit their manuscript. But Hales and Ferguson did not want to spend another year reworking their paper. "Tom could spend the rest of his career simplifying the proof," Ferguson said when they completed their paper. "That doesn't seem like an appropriate use of his time." Hales turned to other challenges, using traditional methods to solve the 2,000-

year-old honeycomb conjecture, which states that of all conceivable tiles of equal area that can be used to cover a floor without leaving any gaps, hexagonal tiles have the shortest perimeter. Ferguson left academia to take a job with the U.S. Department of Defense.

Faced with exhausted referees, the editorial board of the *Annals* decided to publish the paper—but with a cautionary note. The paper will appear with an introduction by the editors stating that proofs of this type, which involve the use of computers to check a large number of mathematical statements, may be impossible to review in full. The matter might have ended there, but for Hales having a note attached to his proof was not satisfactory.

In January 2004 he launched the project "Formal Proof of Kepler," or "FPK" after its initials, and soon dubbed "Flyspeck." Rather than rely on human referees, Hales intends to use computers to verify every step of his proof. The effort will require the collaboration of a core group of about 10 volunteers, who will need to be qualified mathematicians willing to donate the computer time on their machines. The team will write programs to deconstruct each step of the proof, line by line, into a set of axioms that are known to be correct. If every part of the code can be broken down into these axioms, the proof will finally be verified. Those involved see the project as doing more than just validating Hales's proof. Sean McLaughlin, a graduate student at New York University who studied under Hales and has used computer methods to solve other mathematical problems, has already volunteered. "It seems that checking computer-assisted proofs is almost impossible for humans," he says. "With luck, we will be able to show that problems of this size can be subjected to rigorous verification without the need for a referee process." But not everyone shares McLaughlin's enthusiasm. Pierre Deligne, an algebraic geometer at the Institute for Advanced Study, is one of many mathematicians who do not approve of computer-aided proofs. "I believe in a proof if I understand it," he says. For those who side with Deligne, using computers to remove human reviewers from the refereeing process is another step in the wrong direction.

Despite his reservations about the proof, MacPherson does not believe that mathematicians should cut themselves off from computers. Others go further. Freek Wiedijk, of the Catholic University of Nijmegen in the Netherlands, is a pioneer in the use of computers to verify proofs. He thinks that the process could become standard practice in mathematics. "People will look back at the turn of the 20th century and say 'That is when it happened'," Wiedijk says. Whether or not computer checking takes off, it is likely to be several years before Flyspeck produces a result. Hales and McLaughlin are the only confirmed participants, although others have expressed an interest. Hales estimates that the whole process, from crafting the code to running it, is likely to take 20 person-years of work. Only then will Kepler's conjecture become Kepler's theorem, and we will know for sure whether we have been stacking oranges correctly all these years.

13

Has Poincaré's Conjecture Finally Been Solved?

How can an ant determine whether it is sitting on a ball or a bagel? How could the ancient Greeks have known that the earth is not flat? The difficulty with solving problems of this nature lies in the fact that a ball, a sphere with a hole in it, and a flat plane look exactly the same in the neighborhood of the observer.

Topology developed in the 19th century as an offshoot of geometry, but it has become a mathematical discipline in its own right over the years. Topologists study qualitative questions about geometrical objects (surfaces and spheres) in two-, three-, and higher-dimensional spaces. These objects—imagine them made out of clay or Play-Doh—may be transformed into one another by stretching and squeezing but without tearing, piercing, or gluing separate bits together. Spheres or cubes, for example, can be manipulated into egg-shaped forms or pyramids and are therefore considered topologically equivalent. A ball, on the other hand, cannot be transformed into a doughnut without punching a hole in it. Finally, a pretzel is not equivalent to a doughnut, topologically speaking, because it has three holes.

The number of holes that an object sports is an important characteristic in topology. But how does one define holes mathematically? Nothing, surrounded by a border? No, that would not do. Theoretically one can proceed as follows: Sling a rubber band around the object to be investigated. If it is a ball, an egg, or another object without a hole, then no matter where the rubber band was slung, one can always shrink the loop continuously to a single point. But if one takes an object such as a doughnut or pretzel and wraps the rubber band along the surfaces, this does not always work. If the rubber band is threaded through one of the holes, the loop gets caught upon tightening. This is why, in topology, bodies are classified according to the number of their holes.

Surfaces of three-dimensional bodies, such as balls or doughnuts, are called two-dimensional manifolds. What about three-dimensional manifolds—that is, the surfaces belonging to four-dimensional bodies? To investigate these objects, the French mathematician Henri Poincaré (1854–1912) proceeded in the same manner as with two-dimensional manifolds. And he made a bold claim: Three-dimensional manifolds on which any loop can be shrunk to a single point are topologically equivalent to a sphere. When he attempted to provide a proof for this assertion, however, he got into hot water. His attempt proved to be a failure. So in 1904 the word "claim" was changed to "conjecture."

In the course of the second half of the 20th century, mathematicians succeeded in proving that Poincaré's conjecture holds true for four-, five-, six-, and all higher-dimensional manifolds. But the original conjecture for three-dimensional manifolds remained unsolved. This was all the more frustrating, since four-dimensional space, where one finds the three-dimensional manifolds under investigation, represents the space-time continuum in which we live.

In the spring of 2003 it was announced that Grigori Perelman, a Russian mathematician from the Steklov Institute in St. Petersburg, might have been successful in providing a proof for the Poincaré conjecture. Similarly to his famous colleague Andrew Wiles, who had "cracked" Fermat's last theorem in 1995, Perelman too had been working in complete isolation and solitude for eight years. His efforts culminated in three essays that he posted on the Internet, one in November 2002, one in March 2003, and one in July 2003.

Scientists in the former Soviet Union do not have an easy life, and Perelman was no exception. In one of the papers a footnote mentions that he had been able to scrape by financially only thanks to the money he had made as a graduate fellow at American research institutes. In April

2003 Perelman gave a series of lectures in the United States. The intention was to share his results with his colleagues and get their feedback.

In his proof Perelman relied on two tools, developed in previous work by two colleagues. The first is the so-called geometrization conjecture, which was formulated in the 1980s by William Thurston from the University of California, today at Cornell University. It is a known fact—to mathematicians, that is—that three-dimensional manifolds can be decomposed into basic elements. Thurston's conjecture says that these elements can only be of eight different shapes. Proving this conjecture was an even more ambitious undertaking than proving the Poincaré conjecture. The latter only aims to identify the manifolds that are equivalent to a sphere. Thurston managed to prove his conjecture, but only after making certain additional assumptions. For this feat he was awarded the highest mathematical award, the Fields Medal, in 1983. As far as the most general version of the conjecture is concerned, the version without the assumptions made by Thurston, the conjecture remained unproven, however.

The second tool on which Perelman relied is the so-called Ricci flow. This concept was introduced to topology by Richard Hamilton of Columbia University. Basically, the Ricci flow is a differential equation that is related to the dispersion of heat in a body. In topology the Ricci flow describes the development of a manifold that changes continuously, at a rate inversely proportional to the manifold's curvature at every point. This permits a deformed body to find a state of constant curvature. Sometimes the Ricci flow allows manifolds to split into several components. Hamilton proved—albeit also under certain limiting conditions—that these components could only take on the eight shapes predicted by Thurston.

Perelman succeeded in extending the theory of the Ricci flow to a complete proof of the general version of Thurston's geometrization conjecture. From this it follows, as a corollary, that Poincaré's conjecture is correct: If a loop around a three-dimensional manifold can be shrunk to a point, the manifold is equivalent to a sphere.

The proof that Perelman presented during his lecture series still requires in-depth verification. This could take several years. It would not be the first time that a proof has been found to be lacking only after it has been presented to the mathematical community. In 2002, for example, a year before Perelman's publications, the English mathematician Martin Dunwoody posted what he believed was a proof for the Poincaré conjecture on the Internet. (See Chapter 5.) Much to his chagrin, a colleague soon noticed that one claim which Dunwoody had made in the course of his five-page paper was not fully proven.

So far there has been no reason to think that the proof Perelman described in his papers is incorrect. No gaps have been uncovered; no errors have been found. Should his proof pass all future tests, it looks as if the Russian mathematician would be the recipient of the first Clay Prize, to be awarded to mathematicians who manage to solve one of the seven "millennium problems." With the $1 million prize money, Perelman would no longer have to rely on the meager honorariums that guest lecturers generally receive.

IV

Personalities

14

Late Tribute to a Tragic Hero

On August 5, 2002, the world celebrated the bicentenary of the birth of one of history's most eminent mathematicians, the Norwegian Niels Henrik Abel (1802–1829). He died, barely 26 years old, of tuberculosis. Brief though his life was, Abel's work was extremely fruitful. An important encyclopedia for mathematics mentions the name Abel and the adjective "abelian" close to 2,000 times.

The works of Abel are so significant that in 2001 Thorvald Stoltenberg, then Norwegian prime minister, announced the establishment of the Abel Endowment Fund, which would henceforth award an annual prize of 800,000 euros in his name. This prize, which models itself on the Nobel Prize, was meant to become the most significant award in the field of mathematics.

Abel grew up in Gjerstad, a town in the southern part of Norway, as the second oldest in a family of seven children. His father was a Lutheran priest who at one time also acted as a member of the Norwegian parliament. Until the age of 13, Niels was educated at home by his father. It was only when the teenager started to attend a church school in Christiana, 120 miles away, did his talents really became apparent. A mathematics teacher recognized the unusual gifts of the young boy and encouraged him as best he could.

When Abel was 18 years old, his father died and all of a sudden Abel found himself forced to support his family, which he did by offering private lessons in basic mathematics and performing odd jobs. Thanks to the financial help of his teachers, Abel was able, in 1821, to enroll at the University of Christiana, which was later to become the University of Oslo. It did not take long before Abel started to outshine his teachers. His first major achievement, however, proved to be in error. Abel believed that he had found a method to solve equations of fifth degree and submitted his paper for publication to a scientific journal. The editor could not understand the solution, however, and asked Abel for a numerical example.

Abel set out to meet this request but soon became aware of a mistake in the derivation. The error proved to be of some benefit, though. While trying to make the correction, Abel realized that it is simply impossible to solve an equation of fifth or higher degree by means of a formula. To arrive at, and prove, this conclusion Abel made use of a concept called group theory, which would later develop into a very important branch of modern mathematics.

Abel published this paper at his own expense. He then undertook a journey to Germany with the financial support of the Norwegian government in order to call on the famous mathematician Carl Friedrich Gauss in Göttingen. Gauss, however, never read the paper that Abel had sent him ahead of his visit. Furthermore, he let him know in no uncertain terms that he had no interest whatsoever in a meeting. Disappointed, Abel continued his journey on to France, a side trip that had a fortuitous side effect. En route to Paris he made the acquaintance of the engineer August Leopold Crelle in Berlin, who was to become a very close friend and supporter. The *Journal für Reine and Angewandte Mathematik* (*Journal for Pure and Applied Mathematics*), which was founded by Crelle and continues to appear to this very day, published many of Abel's highly original papers.

The French colleagues whom Abel tried to look up proved to be no more hospitable than the German professor. A work on elliptic functions that Abel had produced and sent by way of introduction to Augustin Cauchy, the leading French mathematician of the day, was not even noticed. His paper fell into oblivion and was finally lost altogether. Despite this disappointment, Abel persisted, stayed on in Paris, and did everything he could to gain recognition for his work. He was desperately poor and could only afford a single meal a day.

But in the end none of his sacrifices paid off. Crelle

tried everything to convince his friend to settle in Germany, but Abel, sick and without a penny, returned to his home country. After Abel's departure, Crelle set about to find an academic position for his friend, and at long last his efforts bore fruit. In a letter dated April 8, 1804, he was overjoyed to be able to inform his friend that the University of Berlin was offering him a teaching position as a professor. Unfortunately, it was too late: Niels Henrik Abel had died of tuberculosis two days earlier.

From the many concepts connected to Abel, let us briefly mention the concept of the "abelian group." Modern algebra defines a set of elements as a group if these elements can be connected to each other with the help of an operation. Four conditions must be fulfilled: First, the result of the operation must also be an element of the group. Second, the operation must be "associative," which means that the order in which two successive operations are performed does not matter. Third, a so-called neutral element has to exist, which leaves the result of the operation unchanged. Fourth, each element must have an inverse. For example, the whole numbers form a group under the operation of addition. Here are the reasons why: The sum of two whole numbers is also a whole number; the operation is associative since $(a + b) + c = a + (b + c)$; the number zero is the neutral element because a number plus zero leaves the number unchanged; and the inverse element of, say, 5 is –5. Rational numbers (whole numbers and fractions) do not form a group under multiplication, even though the product of two rational numbers is also a rational number (for example, 2/3 times 3/7 equals 6/21), the inverse element of 5 is 1/5, and the neutral element, in this case, is 1. That is because 0 has no inverse.

Groups can be subdivided into "abelian" and "nonabelian" groups. A group is called abelian if the elements, when connected to one another, can be interchanged (for example: $5 + 7 = 7 + 5$). An example of elements that form a "nonabelian" group are the rotations of dice. If one rotates a die around two different axes in sequence, it certainly matters in which order these rotations take place.

Try it and see for yourself. Take two dice and place them on the table in identical positions. Rotate the first die around the vertical and then around a horizontal axis. Then rotate the second die in the same directions, but first around the horizontal axis and then around the vertical. You will note that the faces of the dice point in different directions. Hence the group of rotations of dice are "nonabelian." It is this particular fact, among others, that makes the solution of Rubik's famous cube so devilishly tricky!

15

The Unpaid Professor

In the year 1934 the sad circumstances in Germany led to good fortune for Zurich. Because of his Jewish background, the mathematician Paul Bernays was forced to move from Göttingen to the city on the river Limmat. His reputation as a brilliant logician preceded him. Born as a Swiss in London in 1888, Bernays first studied engineering, then mathematics, in Berlin. The young doctor then went on to teach for five years as a *Privatdozent*—an unpaid assistant professor—at the University of Zurich.

One day the famous mathematician David Hilbert visited Zurich. During a walk with Swiss colleagues through the hills surrounding the city, he became aware of the talented Bernays and immediately offered him a position in Göttingen. Though the *Privatdozent* was already in his 30s, he did not consider it beneath his honor to move to Göttingen as an assistant to the great Hilbert. The extremely fruitful collaboration culminated in the two volumes of *Foundations of Mathematics*, in which the authors built the edifice of mathematics, based completely on symbolic logic.

But the brown clouds of the Nazi regime had gathered on the horizon. The mathematical teaching staff in Göttingen, which to a good part consisted of men (and a single woman, Emmy Noether) of Jewish faith, were chased away by Hitler's henchmen. Hilbert was dejected about the departure of Bernays, as well as that of all his other Jewish colleagues.

Göttingen's loss was Zurich's gain, since Bernays's arrival seeded the beginning of the flowering of logic in Switzerland. At the *Eidgenössische Technische Hochschule*, he was first a lecturer and then an adjunct professor with half a teaching load. Together with Ferdinand Gonseth and George Polya, Bernays conducted the first seminar in logic during the winter semester of 1939–1940. It was to become a staple on the academic calendar; Bernays organized and led it for decades. Attendance at the seminar was free. Bernays, not being a full-time employee of the university, could have asked the participants for a fee, but then most likely not many students would have come.

Even after his retirement in 1958 and on into old age, Bernays continued to attend this especially lively seminar. A former student remembers standing at the blackboard expounding on a recently published article. He had hardly begun his presentation when Bernays asked the first question, which launched a debate with Professor Hans Läuchli, who stepped to the board and, chalk in hand, tried to solve the problem. Thereupon Professor Ernst Specker got up to present another version. Now Bernays, wanting to give more weight to his point of view, pressed forward. The conversation became more and more lively. The poor student, today a respected professor at the University of Lausanne, could barely, and only with great effort, finish his own presentation.

Bernays passed away on September 18, 1977. After his death, the tradition of logic was kept up by his former colleagues Läuchli and Specker. When Specker retired in 1987, his assistants and students urged him to continue the seminar, which he did for 15 more years. Several participants of the seminar are today professors at universities in different parts of the world.

16

Genius from a Different Planet

A little more than a century ago, on December 28, 1903, one of the most important mathematicians of modern times was born in Budapest, Hungary. John von Neumann is known today as the father of the electronic computer, the founder of game theory, and the pioneer of artificial intelligence. He was also one of the developers of the atom bomb. Von Neumann distinguished himself in traditional areas like pure mathematics and the mathematical foundations of physics, modern topics like computer science, and postmodern topics like neural nets and "cellular automata." (The latter were to be rediscovered a few decades later by another genius, Stephen Wolfram; see Chapter 18.)

Jancsi, as he was affectionately called by his family, was the son of well-to-do Jewish parents. His father, a banker, had acquired the aristocratic title of *von*—to be put in front of the rather pedestrian-sounding *Neumann*—by purchasing it. As was custom among the well-to-do in Budapest, the young boy was brought up by German and French governesses. Already as a child he exhibited signs of genius. He was able to converse in Ancient Greek and could recite by heart whole pages from the telephone directory. It did not take long for Jancsi's teachers at the Protestant high school of Budapest to become aware of his mathematical talents, and they furthered him as best they could. (By the way, von Neumann was not the only outstanding intellect to attend this remarkable school: Eugene Wigner, recipient of the 1963 Nobel Prize for Physics, was a student in the class above von Neumann's, and John Harsanyi, Nobel Prize winner for economics in 1994, also was a graduate, as was Theodor Herzl, the ideological founder of the State of Israel.)

To nobody's surprise, von Neumann was eager to study

mathematics after graduating from high school. But his father thought mathematics was a profession with little prospects and preferred for his son to embark on a business career. Jancsi demurred and, in the end, a compromise was reached, which saw the young von Neumann studying chemistry in Berlin. At least, according to his father, chemistry was a subject that was practical and could provide an income. However, the student simultaneously registered with the faculty of mathematics at the University of Budapest. Needless to say, the strict "*numerus clausus*," aimed at preventing Jewish students from attending the university, in no way hindered von Neumann. He also never experienced any anti-Semitism at the university since he did not attend lectures. He only traveled to Budapest to sit for exams.

In 1923 von Neumann changed universities and moved from Berlin to Zurich, where he enrolled at the prestigious *Eidgenössische Technische Hochschule* (ETH). Besides attending the compulsory chemistry lectures, he participated in the seminars on mathematics offered by the ETH. By 1926 he had gained not only the title of *diplomierter Chemiker ETH* but also a doctorate in mathematics from the University of Budapest. The topic of his thesis was set theory, a field that broke new ground and proved seminal for the development of mathematics.

Shortly afterwards the young *Herr Doktor*—at about that point in time already known in his circles as a genius—arrived in the German town of Göttingen. The town's university contained what was then hailed as the world's leading center of mathematics. Its outstanding representative, David Hilbert, the world's most eminent mathematician of his time, received him graciously. The stay was followed by lecture series in Berlin and Hamburg.

Just before scientists of Jewish origin were banned from European universities, von Neumann traveled to the United States at the invitation of Princeton University. That was in the early 1930s, and quantum mechanics had just been developed by Max Planck, Werner Heisenberg, and others. Von Neumann was able to give the theory what until that time it had lacked—a firm and rigorous mathematical footing. This remarkable achievement earned him an appointment to the Institute for Advanced Study, where he became, together with Albert Einstein, one of the six founding professors. From then until his death, the institute was a true home to the mathematician, who had in the meantime become an American citizen and had changed his name from Jancsi to Johnnie.

Not only was von Neumann interested in the foundations of pure mathematics, he was fascinated by the application of mathematics to other areas. At a time when Europe was swept up by war and when the natural sciences and their application to warfare became increasingly important, his work on hydrodynamics, ballistics, and shock waves aroused the interest of the military. It did not take long for von Neumann to become an adviser to the American army, and from there it was but a small step to his appointment, in 1943, to the Manhattan Project in Los Alamos, New Mexico. Together with a group of Hungarian emigrés, including Eugene Wigner, Edward Teller, and Leo Szilard, he became involved in the development of the atom bomb. (Soon this handful of Hungarian scientists were being referred to by their colleagues as the "people from Mars": With their superhuman intelligence and the incomprehensible language they used among themselves, it was rumored that they must have arrived on earth from a different planet!) In Los Alamos von Neumann came up with the decisive calculations that would enable scientists to develop the plutonium bomb.

In the labs of Los Alamos mathematicians had to solve many tedious, routine calculations. To speed things up, they developed numerical techniques that were carried out manually by dozens of human number crunchers. But the need for something faster became more and more pressing. It so happened that von Neumann was familiar not only with the ideas of Alan Turing, an up-and-coming mathematical talent from Britain who at the time was working on his doctoral thesis at Princeton, but also with those of the engineer John Eckert and his physicist colleague John Mauchly. The former pioneered the ideas of a modern computer; the latter two were in the process of constructing the first American electronic computing device in Philadelphia. Building on their preparatory work, von Neumann went on to develop the ideas that would henceforth be known as "computer architecture." To this day "von Neumann architecture" controls the data flow in every desktop PC. It was von Neumann who recognized that programs could be stored in the computer and called up whenever needed, much the same way as data. Until then the experts had thought it would be necessary to wire programs as a piece of the hardware, the method that had been used in mechanical adding machines.

One day during a discussion with Oskar Morgenstern, an economist who had emigrated from Vienna to Princeton, the two men hit on ideas that would henceforth be known as game theory. Von Neumann and his Viennese friend proved the so-called mini-max theorem, which postulates that for board games it makes no difference whether one maximizes gains or minimizes losses. Game theory, which is applied also in the business world and in international politics, has since developed into a separate scientific branch located somewhere between mathematics and economics. (One of its greatest proponents was John Nash, Nobel Prize winner for economics in 1994—together with John Harsanys and Richard Selten—and the central figure in the movie *A Beautiful Mind*.)

Toward the end of his life von Neumann became interested in the brain. In publications on analogies between the human brain and computers that appeared posthumously, he put forth the opinion that the brain functions in both binary and analog modes. Furthermore, he wrote, it makes use not so much of the von Neumann architecture used in PCs but, rather, methods of parallel processing that are used by today's supercomputers. He thereby anticipated the theory of neural networks, which play an important role in today's artificial intelligence research.

Von Neumann had a keen sense of fun. Together with Marietta, his first wife, and, after his divorce, with his second wife Klari—both of whom hailed from Budapest—he attempted to import to America the cabaret atmosphere that he had gotten to know and enjoy as a student in Berlin. The parties thrown by the von Neumanns in Princeton, which lasted long into the night, became legendary.

Von Neumann was honored during his lifetime with numerous scientific prizes and titles. But his final months were very difficult. At age 52 he learned that he had cancer. He could not cope with the inevitable. The scientist with the restless mind was confined to a wheelchair during the day and suffered panic attacks at night. A year later, on February 8, 1957, Johnnie von Neumann succumbed to his illness at Walter Reed Hospital in Washington, D.C.

17

The Resurrection of Geometry

Back in the 1950s it seemed that geometry as a discipline would soon be all but extinct. School teachers had to teach it to their pupils, of course, but as far as researchers were concerned, this particular branch of mathematics could barely offer them anything of even the slightest interest. All old hat, many mathematicians thought to themselves. Only one among them begged to differ. His name was Harold Scott MacDonald Coxeter.

Coxeter was born in London, England, on February 9, 1907. Already as a young schoolboy in one of London's grammar schools, his astonishing mathematical talent attracted attention. When Coxeter Sr. introduced his son to Bertrand Russell, the philosopher advised that the young boy be tutored privately until old enough to study at Cambridge. Even at Cambridge, one of England's foremost academic centers, it did not take long for Coxeter to gain the reputation of an extremely gifted mathematician. Cambridge's most eminent philosopher, Ludwig Wittgenstein, chose him as one of only five students whom he admitted to his coveted lectures on the philosophy of mathematics.

Coxeter completed his doctoral studies at Cambridge and was then invited to Princeton University as a visiting scholar. Shortly before the outbreak of World War II he accepted a position at the University of Toronto. It was there in this rather remote city, far away from the rest of the world's renowned mathematical centers, that Coxeter worked for over 60 years. Today H. S. M. Coxeter is considered one of the 20th century's most eminent proponents of classic and modern geometry.

In 1938, with Europe and America facing political turmoil, Coxeter was quietly tucked away in his Toronto office, the walls of which were lined with mathematical models. His discoveries and theories went well beyond mathematics, having significant impact on other areas, such as architecture and art. It was Coxeter's work on polyhedra, such as dice and pyramids, and their higher-dimensional counterparts, called polytopes, that paved the way to the discovery of the C_{60} molecule, whose shape resembles that of a soccer ball.[1]

The famous geodesic dome, designed by the American architect Buckminster Fuller and built to mark Expo 67, the World's Fair in Montreal, is one of many constructs that made use of Coxeter's discoveries.[2] One only has to look at how the thousands of triangles that make up this dome are joined together to understand that its actual construction is firmly based on Coxeter's preparatory research.

Coxeter, endowed also with artistic gifts, particularly in music, was fascinated by the sheer beauty of mathematics throughout his life. His collaboration with the Dutch graphic artist M. C. Escher turned into one of the most interesting and fruitful partnerships in history between science and art. By the time he met Coxeter, Escher had become thoroughly bored with routinely placing fish after fish next to bird after bird on a blank canvas. His wanted to do something different. He hoped, in fact, to do nothing less than depict infinity. In 1954 the International Congress of Mathematicians took place in Amsterdam, and it was there that Coxeter and Escher were introduced to one another. The two men formed a friendship that would last a lifetime. Not long after this meeting, Coxeter sent one of his papers on geometry to his new friend so that he could read and perhaps comment on it. Despite his complete lack of mathematical knowledge, Escher was so impressed by the drawings with which Coxeter had illustrated the mathematics that he immediately began to create a set of graphics entitled "Circle Limit I–IV."

---

[1]Coxeter's work enabled the scientists Harold Kroto, Robert Curl, and Richard Smalley to do the research that would eventually win them the Nobel Prize for Chemistry in 1996.

[2]The C_{60} molecule is now known as the "Buckyball" for its resemblance to Buckminster "Bucky" Fuller's geodesic domes.

By allowing shapes to be framed by circles and squares and then gradually decreasing their size as they approach the frames, Escher had achieved his goal: He had depicted infinity.

Coxeter was well aware that he was blessed. He ascribed his long life to his love for mathematics, his vegetarian diet, his daily exercise regime of 50 push-ups in one go, and his great devotion to mathematics. As he told one of his colleagues, he was never bored and got "paid to do what he loved to do."

Coxeter was scheduled to give the plenary address to the Festival of Symmetry that was to be held in August 2003 in Budapest, Hungary. In February the 96-year-old Coxeter, still eager to play his part, wrote to the organizers that he very much would like to attend if indeed he would still be alive at that time. "*Deo volente.*" He would give a lecture on "absolute regularity," he wrote. God wished for things to be different. Only a few weeks after having written the letter, on March 31, 2003, Coxeter passed away quietly at his home in Toronto.

18

God's Gift to Science?

One fine spring day in May 2002, the English-born physicist Stephen Wolfram was at long last ready to present his book, *A New Kind of Science,* to the world. Its publication had been a long time in coming, having been announced, and postponed, several times during the preceding three years. A few months prior to its actual publication a review already hailed the book as a pathbreaking work that would have worldwide significance. This review, incidentally, had, just like the book, been published by Wolfram's own publishing house. If you tend to believe an author's own pronouncements, or take kindly to the media spin put out by public relations firms, then you would have to assume that this book was on a par with Isaac Newton's *Principia* and Charles Darwin's *The Origins of Species*. One may be forgiven for thinking that the author would have no qualms whatsoever in comparing his book to the Holy Bible.

Given the brouhaha surrounding the publication, it was not surprising that *A New Kind of Science* quickly became the Number 1 best-seller at *Amazon.com* and kept this sales rank for a couple of weeks. The five-pound tome comprises no less than 1,197 pages, and they are not easy going. The intrepid reader quickly realizes that Wolfram's intention was nothing less than to turn the entire body of science on its head. In his book Wolfram offers solutions to a broad spectrum of topics, including, to name just a few, the second law of thermodynamics, the complexities of biology, the limitations of mathematics, and the conflict between free will and determinism. In short, the book is hailed as nothing less than the definitive answer to all questions, bar none, with which generations of scientists have been struggling, often in vain. *A New Kind of Science,* the author says in the preface, redefines practically all branches of the sciences. This is what Stephen Wolfram believes or, rather, wants us to believe.

Who is this man who is so convinced of his near-divine gifts? Stephen Wolfram was born in London in 1959 into a well-to-do family. His parents were a professor of philosophy and a novelist. (A note for the chauvinists, if there are any, among the readers: The mother was the professor and the father the novelist.) The young lad was sent to boarding school at Eton. There, at the tender age of 15, Stephen wrote his first paper in physics. It was promptly accepted by a reputable scientific journal. In keeping with the ways of the British academic elite, Wolfram went on to Oxford University, whence he graduated at 17, an age when other boys only just about start writing their university applications. By the age of 20 he had not only gained a Ph.D. from the California Institute of Technology, he could look back on a publication list with close to a dozen articles.

Two years later, in 1981, Wolfram won a fellowship from the MacArthur Foundation, the youngest ever recipient of this award. Commonly called "genius awards," these fellowships, given to exceptional individuals who have shown evidence of original work, allow scientists complete financial independence over the course of five years. For reasons having to do with copyrights and patent laws, the somewhat irascible Wolfram then had a falling-out with CalTech and moved on to the Institute for Advanced Study (IAS) in Princeton, New Jersey. His interests, at the time, spanned cosmology, elementary particles, and computer science.

Eventually he hit on a topic that would provide the foundation for his current and—if you believe his public relations machine—revolutionary discovery: cellular automata. Years earlier, in the 1940s, the legendary John von Neumann, one of Wolfram's predecessors at IAS, had come up with the idea of cellular automata. But that was only in passing and he had soon lost interest in them. Indeed, von Neumann's article on this subject was published only posthumously. Nobody pursued the idea. The topic, in fact, was all but lost.

Then in the 1970s, on the other side of the ocean, John Conway, a mathematician in Cambridge, England, came up with a prototype for cellular automata. It was dressed in the guise of a computer game called "The Game of Life," which actually is not really a game but a concept. "Life" uses a grid that resembles a checkerboard, the difference being that the black and white squares do not necessarily alternate but are distributed randomly. One could interpret these squares as the initial population of a colony of bacteria. A few very simple rules determine how the population procreates. Certain bacteria survive, others die off, and new ones develop.

Despite the fact that the simplest rules—such as "If the bacterium has more than three neighbors, it dies"—determine survival and reproduction, the happenings on the checkerboard are anything but simple. Complex population patterns develop, with some colonies dying off, others appearing seemingly out of nowhere, and still others continuously oscillating between two or more states. Then there are those that simply crumble away until only a few small islands are left. What is so surprising is the fact that a very small number of rules succeed in generating such an enormous number of varying and different phenomena and consequences. When *Scientific American* published an article on "Life," the game became so popular that, according to one estimate, more computer time was spent on it than on any other program. It was all the rage.

Stephen Wolfram was no exception. But he did not simply while away his time with "Life." He went a few steps further. Putting the game through rigorous analytical tests, he cataloged the evolving patterns. His article entitled "Statistical Mechanics of Cellular Automata," published in 1983 in the *Review of Modern Physics,* is to this day considered the standard introductory text to the subject of cellular automata.

Wolfram had in the meantime turned 24 and was still

pursuing an academic career, having moved from the IAS to the University of Illinois. He was convinced that his work was the beginning of what would be renewed public interest in cellular automata. But only very few of his colleagues were interested, and Wolfram—apparently unable to thrive without public accolades and recognition—was quite disappointed. However, never short of ideas, he embarked on a new career. He became an entrepreneur.

Of course, Wolfram did not venture into the field of management unprepared. During his years as a scientist, he had developed a piece of software that does symbolic mathematics: That is, it does not simply perform numerical calculations but manipulates equations, gives the solutions to complicated integrals, and accomplishes a host of other neat things. Within two years he developed it into a commercial product and launched it under the brand name "Mathematica." It became a hit in no time. Today there are approximately 2 million professionals—engineers, mathematicians, and industrialists—throughout universities and large corporations who use "Mathematica." Thanks to this innovative and widely used piece of commercial scientific software, Wolfram, Inc., a firm with around 300 employees, still flourishes today.

Wolfram's newly found fortune gave him the independence to return to scientific work. For the next 10 years he spent each and every night on his research. He was firmly convinced that the grid with its small black squares held the key to the secrets of the universe. Wolfram was convinced he had found the answers to all of life's secrets.

It is not uncommon for natural scientists to believe that all phenomena in physics, biology, psychology, and evolution can be explained through mathematical models. Wolfram was no exception. However, he believes that it is not always possible to explain phenomena by means of formulas, where all one has to do is insert variables and parameters into appropriate slots. Instead, he argued, a series of simple arithmetic computations—so-called algorithms—are repeated over and over and over again. While observing the developing string of intermediate results, it would usually not be possible to predict the outcome. A final result will eventually arise simply by allowing the algorithm to run its course.

By looking at the results of algorithms that simulate cellular automata, Wolfram discovered that they mimic patterns one encounters in nature. For example, the development of certain cellular automata is similar to the development of crystals or to the emergence of turbulence in liquids or to the generation of cracks in material. Hence, he argues, even quite simple computer programs can simulate the characteristics of all kinds of phenomena. From there it is only a small step—for Wolfram, that is—to explain the creation of the world: A few computations, repeated millions or billions of times, would have produced all the complexities of the universe. The thesis, which he develops in the book, is incredibly simple and as such also rather disappointing: Cellular automata can explain *all* of nature's patterns.

In the course of the years of nocturnal research, Wolfram managed to simulate an increasing number of natural phenomena with cellular automata. Sometimes he had to run through millions of versions until he hit on the suitable automaton. But eventually it always worked, whether it was in thermodynamics, quantum mechanics, biology, botany, zoology, or financial markets. Wolfram even asserted that the free will of human beings was the result of a process that can be described by cellular automata. He was convinced that very simple behavioral rules—similar to cellular automata—determine the workings of the neurons in our brains. Seemingly complex thought patterns emerge by repeating these behavior patterns billions of times. Hence, what we have hitherto considered intelligence is, in principle, no more complex than the weather. He also posited that a very simple set of calculations could reproduce our universe to its last and smallest detail. It was only a matter of allowing the algorithm to run for a sufficiently long period of time.

Over the years Wolfram worked entirely on his own, sharing his ideas with only a few trusted colleagues. This had both advantages and disadvantages. On the one hand,

Wolfram did not risk exposing himself to any criticism or, worse still, ridicule. On the other hand, nobody could check his propositions or suggest improvements. But Wolfram did his job well. After reading close to 1,200 pages, even the most skeptical reader will be convinced that cellular automata are able to simulate patterns of innumerable natural phenomena exceedingly well.

But does that mean that automata are indeed at the origin of the patterns observed in nature? No, not by a long stretch of the imagination. It is simply not acceptable to let analogies and simulations act as substitutes for scientific proof. If, for example, we visit Madame Tussaud's Wax Museum and stand in front of a figure that looks exactly like Elvis Presley, must we conclude that Elvis Presley was made of wax? Certainly not. Wolfram would disagree. For him it is all a question of what one demands from a model. In his view a model is good if it is able to describe the most important characteristics of a natural phenomenon. Even mathematical formulas only provide us with descriptions of the observed phenomena and not with their explanations, he argues. If Elvis's most important characteristic was what he looked like, then the wax figure can, for all intents and purposes, be considered a good model, Wolfram claims.

Whether such arguments can convince other scientists remains to be seen. But this is of no concern to Wolfram. He intends for his book to be read by the wide public, not by only the select few, and in this aim he does a very good job indeed—not least because of the well-oiled publicity machine.

19

Vice-President of Imagineering

Daniel Hillis does not look like someone who has just been named the recipient of one of the $1 million "Dan David Prizes," not even remotely. Yet this is precisely what happened to this world-renowned computer scientist and entrepreneur. The award is bestowed each year at the University of Tel Aviv in Israel on a number of scientists for their scientific or technical achievements. Hillis, an unassuming and modest man, is a cutting-edge thinker who numbers among his close friends Nobel Prize winners, famous scientists and professors from the most reputable universities in the United States.

Upon closer inspection, Hillis does not even resemble a researcher, nor for that matter does he look much like an entrepreneur. Sitting opposite him as I interviewed him for the Swiss daily *Neue Zürcher Zeitung* in May 2002, I was reminded of an overgrown child whose mischievous and continuously smiling face has a contagious effect. One instinctively wants to share in his obviously good mood. Comfortably seated in one of the elegant leather armchairs in the VIP lounge of the Hilton Hotel in Tel Aviv, this eminent prize winner, who can take credit for holding no fewer than 40 patents, seems well at ease with himself and the world.

Dressed in jeans, an open-necked shirt, and sneakers, Hillis wears his thinning hair in a ponytail. He has the nonchalant air of a true genius, one of those brilliant individuals who effortlessly come up with the most ingenious ideas while relaxing at a university cafeteria. It is not difficult to picture this unpretentious man as a little boy sitting at home on his bedroom carpet tinkering with his robots. Other comparisons come to mind: Walt Disney's Gyro Gearloose, who amused children all over the world with his impossible inventions, or Q, the mastermind of James Bond's impossible accessories.

But Daniel Hillis is all grown up now. He is neither sitting in his playroom amusing himself with robots nor driving around in a fire engine. He is, in fact, steeped in conversation with Sydney Brenner, distinguished scientist and professor from the Salk Institute for Biological Studies, who happens to be one of the other Dan David Prize laureates. The professor seems to have no problems whatsoever in taking this overgrown child quite seriously.

In fact, just about everybody takes Daniel Hillis seriously. It is not difficult to see why. Hillis was the pathbreaking designer of the legendary "Connection Machine," the computer that incorporated and connected no fewer than 65,536 processors and was thus able to attain unprecedented computational speeds. Working on this computer, Hillis faced problems of enormous proportions, problems that had until then been considered unsolvable. This was because these 65,536 chips, which scientists had until then believed could only run on serial machines, had to be made to run in parallel. Hillis, then still a student at the Massachusetts Institute of Technology (MIT), was inspired by the architecture of the brain when he invented his Connection Machine. But, of course, there are significant differences. On the one hand, the number of chips is still many, many orders of magnitude smaller than the number of neurons in a brain. On the other hand, the speed with which the computer chips communicate with each other is many, many orders of magnitude higher than the speeds with which neurons fire. But Hillis, who had this 65,536-piece orchestra play to the beat of his conductor's baton, knew how to overcome all hurdles. The Connection Machine turned out to be not only commercially feasible but also, conveniently, the topic of his Ph.D. thesis.

One day in 1986, Hillis, the eternal child, felt that it was time to take a break from the daily grind at Thinking Machines, the firm he had founded several years previously in order to develop the Connection Machine. Without further ado, he set off for Disney World in Orlando. Happily settled in front of Snow White's castle, Hillis started writing his doctoral thesis. This was to become a habit. Every day he would go to the theme park, find a quiet spot, make himself comfortable, and write chapter after chapter of his dissertation.

His pioneering ideas on parallel computing met with interest that went well beyond the academic world of computer science. It was above all the commercial world that became intrigued by this invention. Eventually, about 70 of his machines were sold. But it was not smooth sailing all the way. The intricate architecture of the Connection Machine made the task of writing dedicated software very difficult and expensive. And it is well known that without the software the hardware is worth little more than the tin and silicon it is made of. So Hillis decided to seek new avenues of innovation.

His next venture was with Walt Disney Imagineering, where he became vice president of research and development in Mickey Mouse's parent firm, a dream come true for Hillis. Here he could live his childhood dreams to the fullest. Initially, he intended to stay for only two years, but he had so much fun developing innovative technologies for movies, carousels, and TV series that he stayed on for a full five years. However, one morning he woke up to realize that the projects he had been working on were maybe not quite as important and beneficial to mankind as he had originally hoped. Not one to dither, he quickly moved on. But he had learned an important lesson at Walt Disney Imagineering: The value of the art of storytelling for organizing and communicating information cannot be overestimated. Disney's method of conveying information was much more efficient than the methods employed by engineers.

It is not surprising, therefore, that this is what Applied Minds, the firm Hillis subsequently founded, sets itself as one of its main tasks: conveying information to the public in such a way that it can be easily grasped. Together with a filmmaker and a team of some 30 scientists and engineers, Applied Minds is busy inventing "things." Of course, Hillis, as CEO, cannot reveal what exactly the things are that Applied Minds eventually intends to bring to market. This is still a trade secret, he whispers. With a mysterious smile he is only prepared to say that he no longer wants to make computers more intelligent. Now he wants to make humans smarter. Then he adds that the things his firm is working on not only must be significant in terms of what they can actually do but should also be aesthetically pleasing. On top of that they must have the potential to change the world. That's it, plain and simple. The projects he loves most, Hillis reveals, are those that combine hardware, software, and mechanical and electronic problems. He and his firm develop ideas and will build the prototypes. The mundane tasks of production and marketing will then be left to the professionals. Modest words from the ponytailed scientist, who, incidentally, also serves as an adviser to the U.S. government.

What will he do with the prize money awarded to him by the Dan David Prize? Hillis intends to donate part of it to nonprofit organizations, the biggest chunk being earmarked for a foundation that sponsors another one of his designs, the construction of a 10,000-year mechanical clock. Some years ago Hillis, as always nurturing the inner child in himself, played with the bizarre idea of constructing a clock that would run for at least 10,000 years and sound a gong once every millennium. The idea was to encourage people to think long term and to stretch out their sense of time, explains Hillis.

Considering the fact that our civilization is relatively young, the long-term implications of this project are astounding. Will anyone even know what a clock is in 10,000 years or, for that matter, what measuring time implies? Will someone be able to maintain the clock's mechanism or even know how to read the instruction manual? What initially seemed a quite innocent project all of a sudden grew into an undertaking of immense proportions. The planned construction will not only highlight the technological problems that inevitably arise with such a monument but, more importantly, will force the builders and the spectators to focus on issues relating to anthropology, our cultural history, and philosophy.

The first working prototype for this mechanical clock is already up and running and on display in London's Science Museum. Scientists and engineers completed their work on it just a few hours before midnight on December 31, 1999, with barely enough time for Hillis to actually witness the most exciting moment of its construction. He had planned to see with his own eyes how the timing mechanism jumped from 01999 to 02000. But things very nearly went horribly wrong. With only six hours to go to the historic event, one of Hillis's colleagues noticed that the power source for the ring that displayed the centuries was inserted the wrong way. Had the error not been noticed, the most crucial moment of the first millennium of its operation would have been botched: The clock would have moved from 01999 to 02800. Feverishly, the engineers worked until literally the last minute to set things straight. Then, at the stroke of midnight, two deep peals rang out and the date indicator on the clock face changed obediently from 01999 to 02000.

Hillis is thinking of using part of his Dan David Prize money to construct a 10,000-year clock in Jerusalem. He is intrigued by the Holy City's obvious connection to the past and to the future, but possibly even more by the challenge of adding Jewish, Muslim, and Christian calendar systems to it.

Toward the end of the interview Hillis, reverting suddenly to his student days, pulls a writing pad out of his pocket and starts scribbling an equation on to a sheet of paper. It represents a mathematical theorem that, he recounts mischievously, he was able to solve during his MIT days despite the serious doubts expressed by his professor. This professor, a world-renowned expert in combinatorial theory, had to admit that his student was right. A quarter of a century has passed since that day and Hillis, now a million-dollar prize winner seated in the VIP lounge of a luxury hotel, quietly chuckles as he recalls his success back then. The sheer enjoyment of having figured out a difficult mathematical puzzle, and having proved his professor wrong, is still written across his face today.

20

The Demoted Pensioner

Ernst Specker is emeritus professor of mathematics at the Swiss Federal Institute of Technology (*Eidgenössische Technische Hochschule*, or ETH) in Zurich, Switzerland, and just recently celebrated his 82nd birthday. Nevertheless, this sprightly, if somewhat stooped, man is just as quick-witted and lucid as he was in the late 1960s when the author of this book was attending his lectures on linear algebra. The term "emeritus" does not sit comfortably with Specker, however. With healthy common sense he remarks that it only embellishes the true nature of affairs. In fact, he adds with a twinkle in his eye, he rather considers himself as demoted, because whenever he wants to give a lecture or organize a mathematics seminar, he now needs to ask for permission from the university. This remark shows, however, that the demotion some 15 years ago in no way diminished the man's enthusiasm for work. Hardly a week has gone by since his retirement without one of his famous seminars on logic. But one day at the end of the academic year of 2001–2002, it really was over for good: The seminar series, first offered in Zurich some 60 years earlier, was not going to be on the academic calendar any longer. The decision had been made higher up.

Specker is a man of utmost kindness, friendliness, and good humor, a characterization to which a number of former students all over the world can attest from when they had to take their oral examinations at the ETH. A poor examinee, stuck with an incorrect answer and at a loss as to how to proceed, was lucky in such moments to have Specker in the examiner's seat. The *Herr Professor* provided sufficient hints and tips until even the most nervous candidate could not help but stumble over the correct answer.

As a mathematician Specker is one of those enlightened men who was always ready to explore new, even outlandish, ideas. One such case that this reporter can attest to relates to his lectures on linear algebra. The professor would stand in front of the lecture hall and, chalk in hand, expound on systems of linear equations and matrices while busily covering the old-fashioned blackboard with equations and formulas. Before long he would run out of space, erase everything from the board, and repeat the process. So it went, week after week: write, erase, write, erase. Eventually Specker got fed up and sought a different way of presenting the subject to the class. He hit on what he thought was an ingenious idea: After covering the blackboard with white chalk, he resorted to a yellow chalk. Without wiping anything off the board, he simply proceeded to write all over the white equations with the yellow chalk, reminding his students to ignore the white writing and only pay attention to the yellow. As could be expected, within a very brief span of time an unbelievable chaos covered the blackboard. Of course, Specker, being the true mathematician that he is, soon realized that this innovation was unusable. He announced that he was discarding the idea, and an audible sigh of relief could be heard in the lecture hall.

As a youngster Specker suffered from tuberculosis and was forced to spend some of his childhood years in Davos, an Alpine resort in Switzerland known for its dry and clean air. There he studied at a private school before moving to Zurich to attend high school. There was no question in Specker's mind but that he would follow in his father's footsteps and pursue a career in law. Very soon, however, he realized that legal studies did not fulfill him. He was not taken by the manner in which lawyers searched for the truth. The way in which, by contrast, mathematicians sought and provided proof fascinated him. Thus, in 1940, he began his studies at the ETH. In 1949, at the age of 29, Specker was invited to spend a year at the Institute for Advanced Study in Princeton, New Jersey. At this legendary institution Specker got acquainted with such luminaries as Kurt Gödel, Albert Einstein, and John von Neumann.

In the fall of 1950 Specker returned to Switzerland and was immediately offered a position as a lecturer at the ETH; after five years he was appointed full professor. It was then that he made a quite revolutionary discovery: The so-called Axiom of Choice does not hold in the formalized set theory of the Harvard philosopher Willard Van Orman Quine. Not surprisingly, this work caused quite a stir, and Specker promptly received an offer of a professorship from Cornell University in Ithaca, New York.

How does a mathematician get the inspiration for the proof of a problem that has been investigated for years and years? Specker's answer is that one can never predict that. An idea may pop up in one's head, just like that, in the middle of a bath or while shaving. It is important, though—and he lifts a finger to emphasize the point—to be completely relaxed. Tension, he maintains, is counterproductive. And another thing: One should never get frustrated by false starts. False starts, Specker emphasizes, very often contribute to further research or, indeed, even form the basis of it.

Specker declined the Cornell offer since his family preferred to remain in Switzerland, but the fact that he was invited to a prestigious university in America had repercussions back home. Specker's hometown institution, the ETH, realized that they possessed in him an extremely valuable asset which needed looking after. The administration relieved him of the rather tedious introductory lectures that he, like all math professors, had to give to the engineering students. Instead, he was allowed to pursue his own specialties. Over the next 50 years Ernst Specker made revolutionary contributions to the fields of topology, algebra, combinatorial theory, logic, the foundations of mathematics, and the theory of algorithms.

One day the famous Hungarian mathematician Paul Erdös paid a visit to Zurich, and Specker coauthored a short paper with him. This publication earned him the coveted "Erdös number 1." Approximately 500 mathematicians around the world have been accorded this honor. The Erdös number evolved because the Hungarian mathematician collaborated with an unprecedented number of colleagues. To have an Erdös number 1, a mathematician must have published a paper with Erdös. To receive the number 2, he or she must have published with someone who has published with Erdös and so on. Specker, ever the mathematician, immediately translated this into a mathematical formula: Every author publishing a paper with an author possessing the Erdös number n automatically receives the Erdös number $n + 1$. As soon as Specker had become a member of the elite club of mathematicians with Erdös number 1, he immediately found himself the target of a whole slew of mathematicians asking him to coauthor papers. This would, of course, automatically entitle them to the honor of a coveted Erdös number 2. (There are about 4,500 mathematicians with Erdös number 2.)

Specker is frequently asked what logic can offer anybody in day-to-day life. Of course, it can help decide if and when an answer is correct, he says. But there are also other applications, such as linguistics or computer science, which have become rigorous branches of science only after first having been subjected to logical formalization.

A case in point would be the question of whether a computer program exists that can test other programs, as well as itself, as to whether they are correct. Thanks to logic the answer is obvious: No such programs exist. Then there are questions that deal with the complexity of a problem: Is it useful to know, for example, that one is able to decide a given question but that it would take an infinitely long time (or at least a few billion years) to actually calculate the answer? Finally, even issues arising in physics can be tackled by resorting to logical argumentation. For example, together with Simon Kochen from Princeton University, Specker proved, purely by logical argument, that so-called hidden variables cannot exist in quantum mechanics. Thus they cannot explain, as Einstein had hoped, certain quantum mechanical phenomena.

Specker continues to give lectures and to participate at colloquia all over the world, but his family comes first. He loves spending time with his eight grandchildren and fondly recalls a recent lunch he had with one of his granddaughters. He spent a most enjoyable few hours chatting with her about mathematics, and this, he recounts with a happy smile, was "a really wonderful experience."

21

A Grand Master Becomes Permanent Visiting Professor

If you are looking for the grand master of Swiss mathematics, the name Beno Eckmann springs to mind. At 87, Eckmann is permanent visiting professor at the *Eidgenössische Technische Hochshule* (ETH). Despite the emeritus status given him some 20 years ago, he is as active as ever.

Eckmann grew up in the city of Berne, the capital of Switzerland. He was a happy lad who found school life easy enough but enjoyed himself especially during math lessons. As a young boy he showed no signs, however, of wanting to make mathematics his profession. In fact, his tutors advised against this path since, in their minds, everything that could be revealed in the field of mathematics had already been discovered. On top of that, young Eckmann was told, mathematics offered few career prospects.

These warnings notwithstanding, Eckmann decided in 1935 to follow his inclinations and enrolled in physics and mathematics at the ETH in Zurich. Suddenly a new world opened up for him. Here, at one of the most advanced scientific institutes in the world, some of the best-known scientists had taken up teaching positions. Among them were Wolfgang Pauli, the future Nobel Prize winner for physics, and the German mathematician Heinz Hopf. They saw it as their mission to look after the rather small group of mathematics students. Hopf, who had emigrated from Germany in 1931, was at the time the leading mathematician working in topology, then still a very young field that deals with structures of higher-dimensional spaces. Eckmann was aware of the opportunity that presented itself and seized on it with both hands. He asked the eminent mathematician to tutor and guide him in the writing of his doctoral thesis. The thesis was judged to be quite outstanding even when measured against the high standards of the ETH, and Eckman was duly awarded a prize.

Eckmann's reputation soon spread beyond the boundaries of Zurich, and in 1942 he was offered a chair as extraordinary professor at the University of Lausanne, in the French-speaking part of Switzerland. But it was a time of war and Switzerland was under threat. Being a patriot, the young lecturer did not hesitate when he was called up for active service but managed to skillfully combine his army service—as a spotter for the artillery—with his duties at the university. Two weeks of lectures alternated with two weeks of military service.

After the war, Eckmann was invited for a two-year visiting position at the Institute for Advanced Study (IAS) in Princeton, New Jersey. There he got acquainted with Hermann Weyl and other members of what was considered the golden guild of mathematicians and physicists—Albert Einstein, Kurt Gödel, and John von Neumann. Einstein, needless to say, was in huge demand. He was the superstar whom everybody wanted to get to know in person. Actually, the discoverer of relativity was fed up with his celebrity status and the constant stream of visitors. However, Eckmann seemed to be an exception in Einstein's eyes, and the grand old man of physics invited him to his home for tea. Maybe Einstein still had a soft spot for Zurich and Berne in which he had spent some quite memorable years and from where Eckmann himself hailed. More probably the liking he took to the young man from Switzerland was due to his winning personality and his honest and talented approach to science.

The other superstar at the IAS, John von Neumann, was much more approachable, Eckmann recalls. A faint smile settles on his lips when he remembers the anecdotes with which von Neumann regaled his friends back in the Princeton days. (One such story has the mathematician, whose penchant for fast cars did not, unfortunately, match his driving skills, speeding along a country road. "Here I go at 60 miles an hour," von Neumann told his listeners quite seriously, "when, all of a sudden, a tree steps forward and . . . crash.")

In 1948 Eckmann was offered a full professorship at the ETH in Zurich. The list of papers he published adds up to some 120 articles. This may not seem particularly long when compared to what is considered to be a standard "portfolio" for today's mathematicians, but the papers are comprehensive and long. They covered fields that were in constant flux, pointed in new directions, and offered completely new insights.

It is not the list of publications alone, however, that accounts for Eckmann's reputation. What impresses is the number of Ph.D. students he looked after and guided in their work. They number over 60. Doctoral students who chose him as their thesis adviser were particularly impressed by the cutting-edge work in which Eckmann was continuously engaged. They also were attracted by the humane and friendly way in which he communicated with each and every one of them. It is not surprising that having been fortunate to find in Eckmann a model professor, over half of his doctoral students became professors themselves and thus were able to offer proper supervision to their own students. A genealogical tree that hangs on the wall behind this still lean and trim octogenarian lists no less than five generations with over 600 doctoral offspring.

Eckmann was always fascinated with the connections between geometry, algebra, and set theory. He always sought pathways between problems that he had already solved to new mathematical problems. For a research mathematician, relevancy should never be a guiding principle, Eckmann cautions. Nevertheless, sometimes one hits unexpectedly on an application to practical issues. A case in point for Eckmann was a theoretical piece of research that he published in 1954. To his complete surprise, the result found an application to economics nearly half a century later.

Eckmann's influence is also apparent in a project he initiated in 1964. Scientists at the time were at a loss as to how they should disseminate their research. This was before the era of the Internet, and the publication of new results in journals could take many months or even years.

Faster dissemination of new results was possible only occasionally at workshops or conferences. Eckmann was determined to find a solution to this unsatisfactory state of affairs. One day an idea sprang to his mind—how results of broad interest could be published and marketed with little expense. He promptly shared it with one of the heirs of Julius Springer, founder of the famous scientific publishing house of the same name in Heidelberg. The heir in question just happened to be studying biology in Zurich at the time.

The idea was simple enough: Just mimeograph the manuscripts with hardly any editing, bind them, and sell them at the lowest cost possible. Thus the series "Lecture Notes in Mathematics" was born in 1964. It became a most valuable service to the community of mathematicians worldwide and, thanks to the continued supervision and care given to it by Eckmann and one of his colleagues, the series now comprises some 1,800 volumes.

Eckmann never shied away from administrative responsibilities. He still believes that professors owe it to their institutions to devote themselves to administrative duties as well as research work. In this respect he always set a good example. Especially noteworthy is the research institute for mathematics at the ETH (*Forschungsinstitut für Mathematik*), founded in 1964, of which Eckmann was director for a good 20 years. Today many such institutes exist—for example, in Barcelona and in Columbus, Ohio. Eckmann is still associated with a number of such institutes in Israel that he also helped found, namely at the Technion in Haifa, the Hebrew University of Jerusalem, Bar Ilan University in Tel Aviv, and Ben Gurion University in Beersheva.

Reflecting on a mathematical career that has now spanned nearly 70 years, Eckmann cannot but comment on how much his subject has changed. This constant state of flux brings with it not only the necessity but also the opportunity for new approaches and innovative concepts.

Whenever he started exploring a new idea, it was as if he embarked on an adventure. Optimism alternated with disappointment until the moment when, hopefully, a breakthrough was achieved. The feeling that befalls a scientist at such an occasion is impossible to describe, Eckmann says somewhat nostalgically. Only those who have been lucky enough to experience it for themselves really know what this means.

V

Concrete and Abstract Matters

22

Knots and "Unknots"

When Alexander the Great cut the Gordian knot in the year 333 BCE, he almost certainly did not give a thought to the mathematical details of his (mis)deed. Similarly, when it comes to tying knots, neither scouts, mountaineers, fishermen, nor sailors care about the higher mathematics involved in this process. It was only because of an error that scientists were prompted to turn their attention to knots. This is what happened.

Toward the end of his career, the Scottish scientist Lord Kelvin (1824–1907) was of the belief that atoms consisted of fine tubes that would become tied up with each other and then buzz around the ether. Kelvin's theory was generally accepted for about two decades before being proved erroneous. In the meantime, however, this mistaken belief had led Peter Tait (1831–1901), also a Scottish physicist, to categorize all possible knots. (Mathematical knots are different from their more mundane cousins in that both free ends are connected to each other. In other words, knots in knot theory are always closed loops.)

A superficial classification would use the number of crossings of two strands as determinants. This type of categorization does not account for the possibility, however, that two different-looking knots could actually be the same—that is, that one of them could be turned into the other one by picking, plucking, tugging, and pulling at, but without cutting or untying, their strands. Thus if one knot can be "deformed" into the other, the two are identical. Tait discovered this quite intuitively and attempted to account only for truly different knots in his scheme of classification. These so-called prime knots cannot be disassembled into further components.

Tait's classification was not without its errors, though,

as Kenneth Perko, a New York lawyer, discovered in 1974. Working on his living room floor, he managed to turn one knot with 10 crossings into another knot that had been listed as different by Tait.

Nowadays we know that there exists only one knot with three crossings, another one with four crossings, and two with five crossings. Altogether there are 249 knots with up to 10 crossings. Beyond that the number of possibilities rises quickly. There are no less than 1,701,935 different knots with up to 16 crossings.

The central question in mathematical knot theory was and remains whether two knots are different or if one of them can be transformed into the other one without cutting and reattaching the strands. The transformation must be performed by means of three simple manipulations that were discovered by the German mathematician Kurt Reidemeister (1893–1971). A related question asks whether a tuft that looks like a knot is in fact an "unknot" because it can be disentangled by means of the Reidemeister manipulation. The well-known and well-worn trick of miraculously unknotting a complicated-looking tangle of strings that magicians use to their advantage under the astonished oohs and aahs of the audience is obviously based on an "unknot."

Henceforth, mathematicians busied themselves looking for characteristic traits, so-called invariants, which could clearly and unambiguously be attributed to the various knots, thus making them distinguishable from one another. James Alexander (1888–1971), working at the IAS in Princeton, found polynomials (see footnote on page 28) that were suitable for the classification of knots. If the polynomials are different, the corresponding knots are also different. Unfortunately, it soon became apparent that the reverse does not hold true: Different knots may possess identical polynomials. Other mathematicians developed different systems of classification, and others still seek a workable recipe of how to convert identical knots from one form into another, equivalent, form.

Is this really relevant to anybody other than scouts, mountaineers, fishermen, or sailors? Knot theory is an example of a mathematical subdiscipline that was developed before applications were even being considered. But with time, useful implementations of knot theory did surface and knots found applications in real life. Chemists and molecular biologists in particular became interested in knots. For example, some of them study the ways in which the long and stringy forms of the DNA molecule wind and twist so as to fit into the nucleus of a cell. If you were to enlarge a typical cell to the size of a football, the length of a DNA double helix would measure about 200 kilometers. And as everyone knows, long pieces of string have the annoying tendency of spontaneously becoming all twisted and tangled. What scientists are interested in is which forms of knots the DNA strings take on and how they then disentangle again.

And, of course, there are the theoretical physicists. Toward the end of last century it became evident that quantum mechanics and the force of gravity are not compatible. In the 1970s and 1980s quantum physicists suggested "string theory" as a new answer to this puzzle. This theory basically says that elementary particles are tiny little strings, crushed together in higher-dimensional spaces. (So maybe Lord Kelvin's erroneous conjecture wasn't so erroneous after all.) Obviously in this situation too, the strings get entangled, and so knot theory found another application.

There is yet another group of people—scientists included—who are interested in knot theory. It consists of the gentlemen who tie their neckties every morning. Thomas Fink and Yong Mao, two physicists at the Cavendish Laboratories in Cambridge, investigated the ways in which elegant men do up their ties before going to the office in the morning or to a dinner party in the evening. They found that no fewer than 85 different ways exist in which the task could be performed. Not all of them fulfill traditional aesthetic demands, however. There are, you see, a number of issues that need to be considered when performing what is commonly thought of as a rather routine act. For instance, the absolute sine qua non of the elegant knot is that it be symmetrical. Then, as all fashion-conscious men know, only the wider end of the tie may be moved around when tying the knot. And lastly, the number of times one moves the free end either to the right or to the left should be roughly equal. Hence, regrettably, trend-setting gentlemen who would like to adhere to these conditions are unable to take advantage of the full range of 85 possibilities. These unfortunate souls are left with merely 10 different knots to choose from.

23

Knots and Tangles with Real Ropes

For about a century, the mathematical theory of knots has been dealing with "embedding the unit circle into three-dimensional space." A knot is mathematically defined as "a closed, piecewise linear curve in three-dimensional Euclidean space." Mathematical knot theory is a branch of topology that focuses on idealized strings, which are assumed to be infinitely thin. Knot theory not only interests mathematicians but, for once, also fascinates laypeople since it is easy to visualize threads and strings as real objects. The fact that knot theory relates to three dimensions is yet another point in its favor. If one takes knot theory into the realms of four-dimensional space, then all knots—which have been tied using one-dimensional strings—instantly become "unknots."

The physical theory of knots, in contrast to the mathematical version, deals not with infinitely thin abstractions but rather with real ropes that possess a finite diameter or thickness. Scientists dealing with the physical theory of knots are interested, for example, in what sorts of knots can be tied in the real world. Or they investigate how much rope is needed to tie a specific knot. Current thinking is that the length needed to tie a specific knot might be a measure of its complexity. Since stringlike objects, such as DNA, possess finite size, the physical theory of knots can provide much more realistic answers than the abstract mathematical theory to scientific problems.

When dealing with real knots, the actual configuration of the strings is of the utmost importance. In mathematical knot theory, all knots that can be transformed into each other by means of tugging, twisting, and pulling are considered identical. This is not true for the physical theory. Here the exact positioning of the strings is of crucial importance. Any deviation in the arrangement of the pieces of rope, however minute, produces a new knot. In other words, whenever one tugs at a knot, a new knot appears. Each knot has an infinite number of appearances. This difficulty is the reason that seemingly simple problems remain unsolved to this very day.

Take the simplest knot of all, the trefoil or overhand knot. Until very recently nobody knew whether a rope measuring 1 inch in diameter and 1 foot in length could be tied into a trefoil knot. (In knot theory the two ends of the rope must be attached to each other; that is, the rope forms a closed loop. Thus the trefoil knot becomes a cloverleaf knot.)

Simple reflection reveals that a rope which is only π (approximately equal to 3.14) times longer than it is thick does not suffice to tie any knot. One can form no more than a compact ring by linking together the two loose ends. (The length is measured in the middle of the rope.) There is no rope left over for tying the actual knot! Thus π is a lower bound for any knot. Knowing this fact does not answer the question, however, of what the minimum length is to tie a string into a cloverleaf knot. (This is, incidentally, where the builders' cocky retort must derive when asked how long a job will take. "Well, how long is a piece of string?" is the answer, which means, in other words, "Who knows?")

To achieve some progress, knot theorists took recourse to a bright idea. They designed a computer model that depicts knots, with the assumption that repelling energy is distributed along the rope. As a consequence, the strands would be pushed away from each other and the knot would transform itself into a configuration in which the strands are separated from each other as far as possible. Any slack in the strands would become visible and could then be eliminated by pulling and tugging. Based on these and similar exercises, mathematicians continued to seek the minimal length required to tie a knot.

In 1999 four scientists succeeded in calculating a new lower bound for the rope length. They established that even if the rope measures about 7.8 times its thickness

(2.5 times π), it is not sufficiently long to tie a cloverleaf knot. A few years later three other researchers were able to prove that a ratio of length over thickness of 10.7 does not suffice either. It was only in 2003 that Yuanan Diao, a Chinese scientist then at the University of North Carolina, was able to come up with the answer to the original question. It was negative: Diao proved that even 12 inches of a 1-inch-thick rope does not suffice to tie a cloverleaf knot. At the same time he worked out a formula that calculated lower bounds of rope lengths for all knots with up to 1,850 crossings.

Later Diao managed to refine the conditions for a cloverleaf knot even further. He demonstrated that at least 14.5 inches of rope was necessary to tie that particular knot. On the other hand, computer simulations showed that 16.3 inches would suffice to do the job. Obviously, the truth lies somewhere in between these two figures.

Another question with which knot theorists of the physical persuasion grapple refers to the mysterious and complex form of the legendary Gordian knot. Alexander the Great could only undo it with the stroke of his sword. What was its exact configuration? For a long time it had been assumed that the knot was tied when the rope was still wet and that it had then been left to dry in the sun. The knotted rope would have then shrunk to its minimum length. In 2002 the Polish physicist Piotr Pieranski and the biologist Andrzej Stasiak from the University of Lausanne, Switzerland, found such a knot. With the help of computer simulations they were able to create a knot whose rope length was too short to untie. In their statement to the press they said that "the shrunken loop of rope was entangled in such a way that it could not be converted back to its original circle by simple manipulations."

Working on these computer simulations, the two researchers made yet another and wholly unexpected discovery that may turn out to have far-reaching consequences. They defined a "winding number" for knots: Each time one strand of the rope is strung over the other strand from left to right, the number 1 is added. Whenever the strand is strung over the other strand from right to left, the number 1 is subtracted. To their immense surprise, the average winding number—where the average is taken over all viewing angles—amounted to a multiple of the fraction 4/7 for each and every knot they subjected to the calculations. No explanation has so far been offered for this phenomenon. Recall in this context that "string theory" describes elementary particles as small, possibly entangled strings. Hence, some physicists suspect that the quantified properties of elementary particles may rest on this rather mysterious particularity of "knot quanta."

Physical knots find very real applications in daily life, for example, in tying shoelaces. Burkhard Polster, a mathematician at Monash University in Australia, decided to subject this mundane routine to rigorous mathematical analysis. The criteria he used were the length of the lace, the firmness of the binding, and the tightness of the knot. On the assumption that each eyelet of the shoe contributes to the tension of the lacing, Polster proved that the least amount of lace is required when the laces are crossed not every time but every other time. (The precise number of crossings is a function of whether the number of eyelets is an even or odd number.)

It stands to reason that this way of tying up one's shoes does not really provide a lot of firmness. If the tension at the back of the foot is of importance, then the traditional ways of tying the shoes are certainly the best: Either you cross the laces each time they are thread through the eyelets or—another traditional and possibly more elegant way of doing up your shoes—you thread one end of the lace from the bottom eyelet into the top eyelet on the opposite side and thread the other end in parallel strips from one side to the other.

Once the shoe has been laced, what is the recommended way to tie the loose ends into a knot? Most people make a double knot, and the loops serve merely as decoration. But here too things are not as straightforward as one might assume initially. There are, it turns out, two ways of tying this knot, and the difference between them could not be more obvious. One knot is the granny knot, where both ends of the lace are crossed over twice in the same direction. Every boy and girl scout knows that this knot is not sufficiently tight. Proof of this is provided on each and every playground, with mothers forever having to bend down to do it all over again. (No wonder Velcro is so popular. Unfortunately, it deprives children of one of the most exciting learning experiences.) A much tighter and more durable knot is the so-called square knot. It is very similar to the granny knot, with one crucial difference. The knot is first tied by crossing the laces in one direction and then, moving to the second knot, by crossing the two loops in the opposite direction.

24

Small Mistakes May Have Large Consequences

Electronic calculators are precise and never, ever, make mistakes. This at least is what we would like to think. But, in fact, such errors occur all the time. It is just that we hardly ever notice. Take a pocket calculator, for example, which has buttons for "square" and "square root," and follow this procedure: Press the number 10, then the square root button, and then the square button. As expected, the number 10 appears on the display screen, since the square number of the square root of 10 is, of course, 10. So far so good. Now try this: Press the number 10, then press the square root button 25 times, and follow up by pressing the square button 25 times. The result, one would expect, should again be 10, but the display shows something like the number 9.9923974. Ordinarily not much thought is given to this rather minor divergence of 0.07 percent. It is an error one can usually live with. But now repeat the experiment by pressing the square root button and the square button 33 times. The resulting number, 5.5732436 or something similar, no longer bears any resemblance to the real answer, which, of course, is 10.

The reason for this phenomenon, which occurs without fail in one way or another with each and every digital calculator, is the fact that a number can have an infinite number of decimals. An example is the fraction 1/3. Expressing it in decimals results is an infinite number of threes after the decimal point. But—and it is a very big "but"—calculators can only store a finite amount of numbers. As a general rule, numerical values are truncated after 15 digits by computers. Thus very small errors exist between the true numbers and the stored or displayed values.

In general, we just put up with these inaccuracies since it is not difficult to manage daily life with only two or three digits after the decimal point. There are times, though, when rounding errors can lead to catastrophes. On February 25, 1991, during the Persian Gulf War, an American Patriot missile battery in Dharan, Saudi Arabia, failed to intercept an incoming Iraqi Scud missile. The Scud struck an American army barracks and killed 28 soldiers. The cause for this tragic mishap was an inaccurate conversion of time, measured in tenths of seconds, to the binary values as they are stored in the computer. Specifically, elapsed time was measured by the system's internal clock in tenths of a second and stored in binary numbers. Then the result needed to be multiplied by 10 to produce the time in seconds. This calculation was performed using 24 bits. Hence the value 1/10, which has a nonterminating binary expansion, was truncated after 24 bits, resulting in a minute error.[1] This truncation error, when multiplied by the large number giving the time in tenths of a second, led to what was to be a fatal mistake.

On the evening of the election day of April 5, 1992, the Green Party in Germany's state of Schleswig-Holstein was elated. By a hair's breadth, the party had mastered the 5 percent threshold required for entry into the state parliament. The rude awakening came shortly after midnight. The true election results were published, and the Greens discovered to their dismay that they had actually only received 4.97 percent of the vote. The program that calculated the election results throughout the day had only listed one place after the decimal, and the count had been rounded to 5.0 percent. This particular piece of software had been used for years, but nobody had thought of turning off the rounding feature—if not to say bug—at this crucial moment. The long and the short if it was that the Greens were unable to occupy any seat in parliament.

On June 4, 1996, an unmanned Ariane 5 rocket was launched off the island of Courou in French New Guinea, but it exploded just 40 seconds after liftoff. The rocket

---

[1]Bit is short for BInary digiT, the latter being either 0 or 1.

had veered off its flight path and had to be destroyed by ground control. Due to a software error, the guidance system had misinterpreted a rounded figure.

In 1982 the stock market in Vancouver introduced a new index and set the initial value at 1,000 points. After less than two years the index was down by nearly half, even though the average value of the stocks had increased by some 10 percent. The discrepancy was, again, due to rounding errors. While calculating the index, the weighted averages of the stock prices were truncated after too few decimal places.

In one particular instance, however, rounding errors led to a significant discovery. One day in the 1960s, Edward Lorenz, a meteorologist at the Massachusetts Institute of Technology, was busy observing weather simulations on his computer. After a while he felt that he needed a break. Lorenz stopped running the program and jotted down the intermediate results. After finishing his cup of coffee, Lorenz returned to his desk, fed the intermediate results back into the computer, and let the simulation run its course. To his surprise, the weather on his computer took a completely different turn to what he expected based on the previous simulations.

After brooding for a while over this puzzle, Lorenz realized what had happened. Before leaving for the coffee shop, he had copied down the numbers he saw on his computer screen. These numbers were displayed to three places behind the decimal point. But inside the computer, numbers were stored to eight decimal places. Lorenz realized that his computer program had been working with values that had been rounded. Since the weather simulation involved several nonlinear operations, it was not surprising that divergences cropped up rather quickly. Nonlinear expressions—that is, expressions like squaring or taking the square root—have the annoying characteristic of amplifying even minute mistakes very quickly.[2]

---

[2]Linear operators are the arithmetic operations of addition, subtraction, multiplication, and division.

Edward Lorenz's discovery set the foundation for so-called chaos theory, which today is a well-known concept. One of the consequences of this theory is the notorious butterfly effect. Basically it says that the movement of a butterfly's wings may unleash a hurricane at the other end of the world. Tiny vortices in the air, caused by a butterfly's flapping wings may represent no more than a change in the 30th digit behind a decimal point. However, nonlinearities in the weather could augment the tiny air movements a billionfold and thus escalate into a hurricane.

But there is another, less sinister, way of looking at things. By flapping its dainty wings, a butterfly could, by the same token, prevent a hurricane from arising. Mathematical models that make use of the reverse butterfly effect have found applications in, for example, cardiology. Minute electrical shocks, released at precisely the right moment, may correct a chaotic heartbeat and prevent a heart attack.

25

Ignorant Gamblers

Risk is a phenomenon every one of us encounters daily, anywhere we go. However, not everyone knows how to properly deal with its consequences and implications. Just look at the large number of people who fritter away their money in casinos. Do they fail to notice the expensive décor when they set foot in such an establishment, and do they not realize that it is their money that pays for it? Why do so many homeowners refuse to insure their property against earthquakes even though all their belongings are at risk? And why—the height of absurdity—do many citizens insure their belongings against theft and burglary yet have no qualms whatsoever about risking their money week after week on lottery tickets?

One reason people pursue risky activities, be it bungee jumping, delta gliding, or gambling, willingly and without much thought, is that these adventures promise adrenaline surges. Otherwise, people do not spend a great deal of time analyzing the dangers when they embark on such activities. One explanation of why they do, or rather do not, ponder the risks is that statisticians find it very difficult to convey the value of their research to laypeople. So severe does the Royal Statistical Society in England consider the problem that its members decided to devote an entire edition of their journal, *Statistics in Society*, to the topic: How can the wider public be informed of the true extent of the risks involved?

Actually it would be quite simple to calculate the so-called expected values of risky activities and, at the push of a button, obtain the key figure that should give the correct decision. All one has to do is multiply the amount one fears losing by the probability of an inclement event occurring. Unfortunately, more often than not, one or both of the two factors are not easily expressible in numerical terms. For example, how high is the probability that a pedestrian will be hit by a falling flowerpot? And what, in this case, would be the financial damages? Or what value would one put on the life of one's child?

But even in cases where damages and probabilities can be quantified to the last decimal place, most people do not want to take any notice. At roulette, for example, all factors are well known. This in no way fazes devoted gamblers, however. They simply ignore the 2.7 percent probability of the ball falling on the zero. They consider roulette a game of luck in which they have a realistic chance of winning. What gamblers tend to forget is that not only the décor of the casino but also the massive profits going into the pockets of the casino owners are financed out of this little zero.

To the public it does not seem to matter whether an event spells gain or loss. The same attitude holds sway over the average citizen in both circumstances. For example, seismologists in Switzerland have calculated that on average the country will be hit by an earthquake measuring 6 or more on the Richter scale every 120 years. But since no one can predict in which year one such event will take place, the actual probability of such a catastrophe hitting Switzerland amounts to about 0.8 percent every year.

Now, take the value of a one-family home, including its contents, which is, say, $500,000, and multiply this value by the probability of 0.8 percent. Would an annual insurance premium of $4,000 be appropriate? Certainly not, since it is not certain that your home will be completely destroyed even if the earthquake happens to be a particularly big one. The relevant question that arises is whether the earthquake of the century will render one in 10 houses or one in a 100 uninhabitable? Assuming the latter case, an annual insurance premium of $40 would be appropriate and fair.

Pessimists who fear that the next earthquake is overdue because the last quake struck in 1855 would, however, be as mistaken as the optimists who opine that surely nothing will happen next year since nothing has happened since time immemorial. These people belong to the category of oddballs that includes gamblers who believe the outcome of the next spin at the roulette wheel must be red because the ball has fallen on black eight times consecutively.

In public life, much more important and far-reaching decisions need to be made than in private life, where one may merely have to decide whether or not to buy an insurance policy. It is a sad fact, unfortunately, that even politicians do not pay close attention to statistical cost-benefit analyses. Is the expected risk emanating from an atomic plant really so much larger than the number of dead and injured that the mere construction and maintenance of a coal mine or dam might claim? And when President Reagan decided to invest $9 million on research into Legionnaire's disease, as opposed to only $1 million for medical research on the AIDS virus, was it possible that his decision was influenced by a prejudice against homosexuals?

The truth is that politicians, like anyone else, allow themselves to be swayed by public opinion. Having the coast guard mount a search and rescue operation with large numbers of helicopters and lifeboats to rescue a shipwrecked fisherman earns the politician of that state many more votes than, for example, straightening out a dangerous curve in the road. In Switzerland it is no different. Huge amounts of rescue equipment are available and ready for use to save the odd mountaineers caught in glacier crevasses. At the same time, dozens of pedestrians may get killed every year in a city when crossing roads simply because there is no budget available to construct overpasses. But it is not always politically correct to ask poignant questions about raw costs and benefits. After all, the Alps are a national treasure to the Swiss. Their accessibility must be safeguarded, costs be damned. All that statisticians can do is provide politicians and managers with the necessary information. It is up to the latter to make the correct decisions.

26

Tetris Is Hard

For over 15 years, millions of people have spent and wasted their time on the computer game Tetris, in which players must try to place different-shaped bricks that float down the computer screen on to a board. The game's aim is to cover the board leaving the smallest number of gaps, which can be achieved by rotating bricks or moving them sideways until the screen is filled all the way to the top.

A group of computer scientists at the Massachusetts Institute of Technology (MIT) discovered that Tetris is far more than just a fascinating computer game, however. In October 2002, Eric Demaine, Susan Hohenberger, and David Liben-Nowell proved that Tetris belongs to a well-known class of problems whose solutions require huge amounts of computer time. The most famous of them is the "traveling salesman problem." A salesman is required to visit a number of cities and wants to do so by the shortest route, without stopping off in any one city more than once. This problem can be solved by computer, but as it turns out, the time required to calculate the route grows exponentially as the number of cities is increased. For that reason this problem belongs to the class of so-called NP problems. These problems differ from P problems, whose calculation time grows at a much slower rate. Problems are said to belong to the P class if they can be solved in an amount of time that is proportional to a polynomial (hence the letter P).

In theory, NP problems could also be solved in polynomial time. But for this to actually happen, one would need a so-called nondeterministic machine (hence the name of the problem: NP stands for *n*ondeterministic *p*olynomial). But such machines, for example, the much-vaunted quantum computers, do not exist and possibly never will. Hence, computer scientists still search for algorithms that can solve NP problems in polynomial time. (Is it possible that such algorithms exist but have simply not been found yet? Or maybe the CIA, the M15, or the Mossad already use them to decipher codes and just don't let on?)

In the meantime, some consolation might be gained from the fact that when tackling an NP problem, one can at least verify possible solutions in polynomial time. For example, *seeking* the prime factors of 829,348,951 belongs to the class of NP problems. But *verifying* that 7,919 is one of the prime factors is only a P problem. All one has to do is divide the larger number by the smaller number and verify that nothing is left over. This is possible in polynomial time.

The first theoretical advance toward answering the above question was made when Stephen Cook, a computer scientist at the University of Toronto, proved in 1971 that all NP problems are mathematically equivalent to each other. This means that if only one NP problem can be solved in polynomial time, all NP problems can be solved in polynomial time. This would imply that all NP problems belong to the P class, a relationship that computer scientists express in the concise formula P = NP. Whether this equation will hold remains an open question. Numerous scientists have wrestled with it, not least because the Clay Foundation has offered a $1 million prize for a correct solution.

Today's computer scientists are still a long way from solving the P versus NP problem. In the meantime they keep themselves occupied with somewhat less challenging problems, such as Tetris. What the MIT researchers discovered was that Tetris is an NP problem. They proved it by reducing the game to the so-called three-partition problem, which has been known since 1979 to be an NP problem.

In this problem a set of numbers must be separated into three separate groups such that the sums of all groups are equal. In their proof, Demaine, Hohenberger, and Liben-Nowell started off with a very complex Tetris position. They proved that, starting with this situation, filling the game board is tantamount to the solution of a three-partition problem. Thus Tetris belongs to the long list of NP problems, as does also, for example, the game Minesweeper that is bundled together with Microsoft Windows. Proof that Minesweeper belongs to the class of NP problems was provided in the year 2000, by Richard Kaye from the University of Birmingham in England.

This does not in any way, however, bring us any closer to the solution of the fundamental problem. Only when an algorithm that detects mines in Minesweeper or fills the Tetris board in polynomial time is found can the $1 million prize be claimed. For now the question remains:

P = NP?

27

Groups, Monster Groups, and Baby Monsters

Algebraic groups consist of elements, such as the whole numbers (–3, –2, –1, 0, 1, 2, 3 . . .), and an operation. The operation—for example, addition—combines two elements. In order for the elements to form a group, it is necessary that the combination of two elements also belong to the group, that the order in which two successive operations are performed does not matter, that the group contain a neutral element, and that for each element there exists an inverse element. Whole numbers, therefore, form a group "under addition." So do the even numbers, since thc sum of two even numbers is also an even number and since the inverse value of, say, 4 is –4. In both of these cases the number 0 forms the neutral element, since any number plus 0 leaves the original number unchanged. Odd numbers, however, do not constitute a group under addition because the sum of two odd numbers is not an odd number.

Whole numbers and even numbers are groups that contain an infinite number of elements. But there are also small groups that only contain a finite number of elements. One example would be the "clock face group." This group contains the numbers 1 to 12. If you choose the number 9 in this group and add 8 to it, the result displayed on the clock will be 5 (12 in this case is the neutral element, since adding 12 to any other number gives the same number).

One of the most significant mathematical achievements of the 20th century was the classification of all finite groups—an accomplishment that in terms of its importance can be compared to the decoding of DNA or to the development of a taxonomy for the animal kingdom by Carl von Linné in the 18th century. To accomplish this truly mammoth task required the united efforts of dozens of mathematicians all over the world.

In 1982 the American mathematician Dan Gorenstein was able to declare victory in the battle for the classification of all finite groups. Gorenstein had been coordinating the worldwide efforts of group theorists. No fewer than 500 publications, comprising some 15,000 printed pages, were needed to prove that there exist exactly 18 families of finite simple groups and 26 groups of a different sort. Small wonder this theorem was dubbed the "enormous theorem."

Back in the 1960s, most experts thought it would take until far into the 21st century before the work would be completed. Some very unusual groups had been discovered that could not be classified into any of the schemes that had so far been developed. They were called "sporadic simple groups"—sporadic because they were rare and simple because . . . no, there is nothing that could be considered simple in the usual sense.

At about that time, John Leech, a mathematician at Glasgow University in Scotland, was studying so-called high-dimensional lattices. A mathematical lattice can be envisioned as a wire mesh. The wire netting around a tennis court is a lattice in two dimensions. The climbing frame in a playground is a lattice in three dimensions. Three-dimensional lattices play a very important role in crystallography—for example, where they illustrate the physical arrangements of the atoms. But Leech was not satisfied with two or three dimensions. He had discovered a 24-dimensional lattice, that would, from then on, carry his name: the Leech lattice. He set about to investigate its properties.

The most important characteristic of a geometric body is its symmetry. Just as a symmetric die looks exactly the same after it has been turned around on any of its axes, a Leech lattice too can be twisted, turned, and flipped—albeit in 24 dimensions—and always remain similar to itself. If a body has more than one symmetry, it is possible to rotate it around one axis, then another one, then around the first one but in the opposite direction, and so on. Since the body is symmetrical, it will look the same after each of the rotations. From this it follows that one can "add" rotations—by performing one after the other—without changing the perceived shape of the body. Furthermore, it is possible to reverse a rotation by turning the body around the axis in the other direction.

The facts that symmetries can be added and that to each rotation there exists an inverse rotation are precisely the requirements that define them as a group. (The neutral element is the "no rotation" rotation.) Hence, the rotations of a symmetrical body can be thought of as the elements of a group. The actual properties of the group depend on the particular body itself.

This is one of many instances in which different mathematical disciplines—in this case geometry and algebra—meet up. Henceforth, mathematicians could cope with geometric problems in the area of symmetry simply by making use of algebraic tools. Leech suspected that the symmetry group of his lattice was of considerable interest, but soon realized that he did not have the group theory skills necessary to analyze it. He tried to get others interested in the question but couldn't. Finally, he turned to a young colleague in Cambridge, John H. Conway.

Conway was the son of a school teacher and grew up in Liverpool. He was awarded his doctorate in Cambridge and was appointed lecturer in pure mathematics. Very soon, however, he started to suffer from depression and came close to a breakdown. He was unable to publish any research results. It was not that he doubted his competence, but how could he convince the world of his abilities unless he published? Leech's lattice appeared at just the right time. It would turn out to be his life saver.

Conway was not a very well-to-do man and, to help supplement his meager income, the depressed mathematician had to take on students for private tuition. Understandably this left him with little time for research and hardly any time for his family. But the opportunity Leech offered him was the stepping stone for which the young Cambridge mathematician had long been hoping. He would not let it pass him by. Over dinner one evening Conway explained to his wife that during the coming weeks he would be preoccupied with a very complex and very significant problem. He would be working on it every Wednesday from 6 p.m. until midnight and then again every Saturday from noon until midnight. But then, to Conway's utter surprise, he solved the problem in no more than a single Saturday session. On that very afternoon Conway discovered that the group describing Leech's lattice was nothing less than a hitherto undiscovered sporadic group.

The Conway group, as it was referred to from then on, contains a gigantic number of elements: 8,315,553,613, 086,720,000 to be exact. The mathematical community was taken completely by surprise by Conway's breakthrough, which advanced worldwide efforts to classify finite groups a considerable distance. More importantly for Conway himself, the discovery boosted his self-confidence and transformed his mathematical career. He was elected a member of the Royal Society and has remained since then at the cutting edge of research in mathematics. In 1986 he accepted an appointment to Princeton University.

As an aside it should be noted that the Conway group is by far not the largest sporadic group. Still to come at the time was the so-called monster group, which was discovered in 1980 by Robert Griess from the University of Michigan. It contains close to 10^{54} elements, thus possessing more elements than the universe contains particles. The monster group describes the symmetries of a lattice in 196,883-dimensional space. And then there is the so-called baby monster. Containing a "mere" 4×10^{33} elements, it is still a few shoe sizes larger than the Conway group. Even mathematicians who do not easily lose their cool over curious objects find sporadic simple groups unusually bizarre.

28

Fermat's Incorrect Conjecture

Sometimes work in one discipline of pure mathematics has a completely unexpected payoff in another. Some of the famous mathematician Pierre de Fermat's (1601–1665) work in number theory bears this out. It took 150 years, however, until the mathematician Carl Friedrich Gauss (1777–1855) found a geometric application to one of Fermat's statements in number theory: the construction of regular polygons by ruler and compass.

Fermat is famous not only for his notorious "Last Theorem," which had actually been no more than a conjecture until it was finally proven by Andrew Wiles in 1994. As a magistrate in the French city of Toulouse, a position he held throughout his adult life, Fermat was apparently not being kept too busy. His post left him sufficient time to pursue his mathematical passions. In correspondence with Marin Mersenne, a monk who shared Fermat's love for mathematics, he discussed problems in number theory. Mersenne was preoccupied at the time with numbers of the form $2^n + 1$, and Fermat conjectured that, if n is a power of 2, the numbers are always prime. Ever since, numbers of the form $2^{2^n} + 1$ are called Fermat numbers.

Fermat himself did not provide any proof for this conjecture. (In fact, many of his proofs are lost, and it is possible that some of them were not rigorous. Deductions by analogy and the intuition of genius sufficed to lead him to correct results.) Of the Fermat number he knew only the zeroth and the four subsequent ones (3, 5, 17, 257, 65537). The next Fermat number, $2^{32} + 1$, was too large to be computed in his time, let alone be checked for primality. But the first five Fermat numbers are, indeed, divisible only by 1 and by themselves. To conclude from this that all such numbers are prime would have been a rather daring leap, however. And indeed, this was one case where one of Fermat's conjectures was wrong, which may have a rather sobering effect on us mortals: Not all conjectures made by famous mathematicians are necessarily correct.

It was nearly a century, however, before Leonhard Euler of Basle, Switzerland, provided a counterexample. In 1732 he showed that the Fermat number corresponding to $n = 5$ (which equals 4,294,967,297) is the product of the numbers 641 and 6,700,417. Hence not all Fermat numbers are prime. But which are and which are not?

The search for the answer is still going strong. In 1970 it was shown that the Fermat number corresponding to $n = 6$ is also a composite number. Today, volunteers all over the world offer their PCs' idle time to test for the primality of Fermat numbers. In October 2003 it was announced that the Fermat number $2^{2^{2,478,782}} + 1$—a number so unimaginably huge that writing it down would require a blackboard thousands of light years long—is composite.

Unfortunately, there are large gaps in the list of tested numbers. In fact, a mere 217 of the first 2.5 million Fermat numbers have been tested so far. Contrary to Fermat's prediction, not one of them—with the exception of the first five—has turned out to be prime. The failure to come up with any more Fermat numbers that are prime led to a new conjecture—the direct opposite of Fermat's original one: All Fermat numbers, excepting the first five, are composite. This new conjecture remains, like the old one, without proof. Nobody really knows whether there are more than five prime Fermat numbers, if there exist an infinite number of composite Fermat numbers, or if all Fermat numbers, excepting the first five, are prime.

And now to the geometric application.

In 1796 Carl Friedrich Gauss, a 19-year-old student at the University of Göttingen, was thinking about which regular polygons could be constructed using only ruler and compass. Euclid had, of course, been able to construct regular triangles, squares, and pentagons already. But 2,000 years later mankind had not gotten much further than that. What about the regular 17-cornered polygon? To his immense satisfaction, the young Gauss was able to prove that the so-called heptadecagon can actually be constructed. But he did still more. Gauss proved that every polygon whose number of corners is equal either to a prime Fermat number or to the product of prime Fermat numbers is constructible by ruler and compass. (To be precise, this also holds when the number of corners is doubled and redoubled, since angles can always be halved with the help of ruler and compass.)

It follows that it is also possible to construct a 257-cornered polygon, corresponding to the next Fermat number, and instructions as to how to go about constructing a regular 65,537-cornered polygon, on which a certain Johann Gustav Hermes spent 10 years of his life, are tucked away in a box that today is safeguarded in the library of the University of Göttingen.

Gauss suspected that this statement also holds in the opposite direction: The number of corners of any polygon that is constructible by ruler and compass must be a product of Fermat numbers. Indeed, this time the conjecture was correct. It was not Gauss, however, who proved this. The honor fell to Pierre Laurent Wantzel, a French mathematician, who presented a proof in 1837.

During his lifetime, Gauss made innumerable mathematical discoveries. Yet he considered the one involving the construction of the 17-cornered polygon one of the most significant. So highly did he value the discovery of his youth that he expressed the wish to have an image of this figure carved onto his tombstone. The stonemason, thus goes the story, refused, claiming that a 17-cornered polygon would be much too close to a circle. Eventually, a monument was commissioned by the city of Brunswick, Gauss's birthplace, whose pillar was decorated by a 17-pointed star.

29

The Crash of Catastrophe Theory

Every year natural disasters cause damages running into the billions of dollars. A mathematical theory that would help explain, predict, and even avert these fateful phenomena would certainly go a long way in allaying our fears and reducing damages. Such a theory was, in fact, developed some 30 years ago. Unfortunately, it did not live up to the expectations that had been placed in it. So-called catastrophe theory experienced its short yet brilliant rise to fame in the 1970s and 1980s, shortly before disappearing with hardly a trace. The theory nevertheless deserves to be taken seriously. It not only deals with disasters in the conventional sense of the word but shows how sudden and abrupt changes in nature may occur even though the underlying parameters are being varied only gradually.

Phenomena related to catastrophe theory can easily be observed in daily life. Take, for example, a kettle of water on the kitchen stove that is slowly heating up. The water starts to bubble with increasing intensity, until suddenly—at precisely 100 degrees Celsius—something quite different happens: It starts to boil. The water becomes gaseous, that is, it starts to evaporate—an effect known in physics as a change of state.

Another area in which catastrophe theory can be applied is the stability of structures. The more weight that is loaded on to a bridge, the more it suffers a certain amount of deformation. The changes are usually barely perceptible. But at a certain point, disaster strikes: The bridge collapses. For these and other catastrophes only very few variables are responsible. Over vast ranges of values, changes in these so-called control variables do not entail any observable reaction. But as soon as one of them moves ever so slightly beyond the critical point, the catastrophe occurs. It is the straw that breaks the camel's back.

Catastrophe theory was developed by the French scientist René Thom, who died on October 25, 2002. Born in 1923 in Montbéliard in eastern France, he spent the first few years of the Second World War with his brother in safety in Switzerland before returning to France. In Paris he attended *Ecole Normale Supérieure,* an elite college for the very best students, from 1943 to 1946. He was hardly an eminent mathematician at the time. It was only on his second attempt that he managed to pass the grueling entrance exam to *Normale Sup.* But not long thereafter Thom wrote a brilliant doctoral thesis for which he received in 1958 the most significant award a mathematician can get, the Fields Medal.

Some years later Thom succeeded in proving a surprising theorem. He was attempting to classify "discontinuities" and discovered, to his utter surprise, that such breaks in continuity could be classified into a mere seven categories. This astonishing result showed that a whole variety of natural phenomena could be reduced to a mere handful of scenarios.

Thom called the discontinuities "catastrophes" and—as every public relations professional knows—the name is everything. Henceforth catastrophe theory was in everyone's mouth. Unfortunately, at some point, Thom's theory fell into the wrong hands. His main book on the subject—quite incomprehensible to the layman—became a best-seller that many people were quite happy to put on their bookshelves without ever having opened it.

Then the members of other disciplines started to take notice. True, Thom himself suspected that the results of his research could be applied to disciplines other than physics. But when social scientists and other representatives of the "soft" sciences, who normally do not work quantitatively, started to become interested in Thom's new theory, there was no turning back.

The respectability of catastrophe theory was irretrievably lost. All of a sudden and at every turn people thought that they detected Thom's catastrophes. Psychologists diagnosed them in the sudden irate outbursts of choleric patients, linguists detected them in sound shifts, behavioral scientists saw them in the aggressive behavior of dogs, financial analysts detected them in stock market crashes. Sociologists interpreted prison revolts and strikes as Thom catastrophes, whereas historians thought that revolutions fell into that category and traffic engineers believed the same to be true for traffic jams. Even Salvador Dali was inspired by catastrophe theory for one of his paintings.

At the beginning mathematicians were delighted by the fact that at long last their discipline was being recognized by colleagues in other fields. But it did not end well. Experts, or at least those who thought of themselves as such, believed they could predict the exact timing of such discontinuities. The ability of foretelling the precise moment of the next stock market crash or the time of the next outbreak of civil war would, they thought, only be a matter of time. But matters had gone too far. In 1978 the mathematicians Hector Sussmann and Raphael Zahler published a devastating critique in the philosophical periodical *Synthèse*. They attacked the failed attempts of those who sought to apply catastrophe theory to social and biological phenomena. They argued that the mathematical theory only had a right to exist in the fields of physics and engineering.

Then one day the theory was gone. It had disappeared from the scholarly literature, just as suddenly as the catastrophes it claimed to analyze. Maybe it would have been better had the theory not become so popular and if it had, instead, oriented itself along the doctrines of the Kabbalah, the esoteric teachings of Judaism. Since Kabbala is a secret and mystical school of thought, its teachings may only be transmitted to mature men, so that overzealous laypeople would not be able to wreak havoc with it. Such an approach would have surely been of benefit to catastrophe theory.

30

Deceptive Simplicity

Most children are able to deal with integer numbers in kindergarten. Manipulating fractions is a bit more difficult. The dear little ones will have had to attend primary school for a couple of years in order to handle them. But irrational numbers are a different matter altogether. Dealing with numbers that cannot be expressed as a fraction of two whole numbers is where the real difficulties start.

The exact opposite holds true for equations. It is fairly easy to find irrational solutions to problems. The real rub starts when a problem requires that the solutions be only integer numbers. The section of mathematics that deals with such problems is called number theory. An annoying characteristic of the discipline is its apparent simplicity. At first glance, statement of the problems seems easy enough. It is only when one delves into it more deeply that the horrendous difficulties become apparent.

The Greek mathematician Diophantus, who lived some 1,800 years ago in Alexandria and is often known as the father of algebra, is said to have founded number theory. In his honor, equations with unknowns that must be integer numbers are called Diophantine equations.

Diophantus' main work, the *Arithmetika*, consisted of some 130 problems and their solutions. Unfortunately, the books were destroyed during a fire in the small library of Alexandria in the year 391. Many years later, in the 15th century, six of the original 13 books were discovered. (Then, in 1968, another four volumes surfaced, albeit in an incomplete Arabic translation.) For years people puzzled over the manuscripts of the ancient Greek mathematician and only in the 17th century was someone finally able to get a handle on the material. This man was Pierre de Fermat, a French magistrate who enjoyed spending his spare time playing around with mathematics. Today Fermat is known above all for his notorious Last Theorem. (See also Chapter 28.)

One problem that originated with Diophantus remains unsolved to this very day: Which numbers can be expressed as the sum of two integer numbers or fractions, each of them raised to the third power? The question can be answered in the affirmative with the numbers 7 and 13, for example, since $7 = 2^3 + (-1)^3$ and $13 = (7/3)^3 + (2/3)^3$. But what about numbers like 5 or 35? To answer the question, one needs to be familiar with the most complicated methods in modern mathematics.

All that mathematicians have found so far is a method to determine whether a decomposition of a specific number exists. But they are unable to provide the decompositions themselves. To determine whether a number can be decomposed into cubes, the graph of a so-called *L* function must be calculated for this number. If the graph intersects or touches the *x* axis of the coordinate system at precisely the point $x = 1$, the number in question can be decomposed into cubes. If the value of the function at $x = 1$ is not 0, no decomposition is possible. The condition is met for the number 35: The *L* function associated with it becomes 0 at exactly $x = 1$. Indeed, 35 can be decomposed into $3^3 + 2^3$. On the other hand, for the number 5, the graph of the *L* function neither touches nor intersects the *x* axis. This proves that 5 cannot be decomposed into cubes.

Don Zagier, the director of the Max Planck Institute of Mathematics in Bonn, Germany, gave two lectures to the general public in 2003 in Vienna about Diophantine cubic decompositions. Zagier is one of the world's leading mathematicians, and his main area of work is number theory. As a child he was already known to be a wunderkind. Born in the German city of Heidelberg in 1951, he grew up in the United States, finished high school at age 13, completed undergraduate studies in mathematics and physics at the Massachusetts Institute of Technology at 16, and gained a Ph.D. from Oxford at 19. By the age of 23 he obtained the habilitation—the German qualification to teach as a professor—at the Max Planck

Institute of Mathematics. At the age of 24 he was the youngest professor in all of Germany. His talents are not limited to mathematics, by the way: He speaks nine languages.

One of Zagier's talks, part of the Gödel lecture series in Vienna, was entitled "Pearls of Number Theory." The other lecture was held at the opening of "math.space," a unique hall in Vienna's museum quarter whose purpose it is to host popular presentations on mathematics. The hope is that this allegedly esoteric topic could be made accessible to the city's wider public who usually pass their time at operas and in coffeehouses.

Zagier is a quirky little man. But when he starts explaining his pet theory to the audience his performance would make a rock star pale in envy. Constantly jumping back and forth between two overhead projectors, he enthralls his audience with mathematical explanations, delivered in perfect German albeit tinged with an American accent. Even a fierce mathematophobe would forget that he is listening to a math lecture. The joy that Zagier—known by some as Bonn's superbrain—feels for his vocation is obvious to all. Watching him is like watching a concert virtuoso. It is hard to believe that mathematicians, such as Zagier, are sometimes accused of being involved in a dry sort of science.

31

The Beauty of Dissymmetry

Symmetrical symbols, drawings, and buildings have fascinated men and women for millennia. In prehistoric times, craftsmen created symmetrical pieces of jewelry, inspired possibly by the human body and the animal body. One of the oldest symmetrical objets d'art crafted by human hand is a bangle, decorated with an extremely complicated drawing, that was found in the Ukraine and dates from the 11th millennium BCE. Ancient architecture also boasts examples of symmetry, such as the pyramids of Giza (3000 BCE) and the arrangement of rocks at Stonehenge (2000 BCE).

But symmetry does not belong exclusively to the domain of art, nor does it belong only to architecture. Scientists also claim ownership. And once scientists set to work, it is usually mathematics that provides the language and furnishes the tools to explore nature's phenomena.

In elementary geometry there are three broad sorts of symmetry. First, there are figures, such as the letters M or W that are "reflection symmetric": The two halves are mirror images of each another. The line separating the halves—the vertical line through the middle of each letter—is called the axis of symmetry. Second, there are shapes, such as the letters S or Z, that have rotational symmetry. They coincide with themselves after having been rotated 180 degrees about some point. The point about which they are rotated is called the center of rotational symmetry. Finally, an infinite sequence of shapes or signs, such as KKKKKKKKK or QQQQQQQQ, is said to be translation symmetric, since the pattern coincides with itself when slid (translated) to the right or to the left. More complicated types of symmetry also exist, and different symmetries can be combined with each other.

Wallpaper patterns, for example, can have reflection, rotational, and translation symmetry all at the same time.

During the summer of 2003 a conference entitled "Symmetry Festival" took place in Budapest at which scientists and artists gathered for a week of interdisciplinary discussions. They scrutinized examples of symmetry, among them batik weaving, the Talamana system of proportions in Indian sculptures, and, a staple of this kind of art, the pictures of M. C. Escher. This was also an occasion to dispel, once and for all, some of the myths that have become so dear to us all. It turns out, for example, that the pentagram, the five-pointed figure widely believed to have been used by the Pythagoreans as a secret sign, did not in fact have this significance at all. The only source attributing the pentagram to the Pythagoreans dates from the 2nd century AD, 700 years after the death of Pythagoras. More credible sources identify this sign to be the seal of King Solomon. It then mutated into the six-pointed Star of David, which today decorates the national flag of Israel. The myths surrounding the golden section, or "the divine proportions," apparently also belong to the realm of fantasy and fable. It came to be regarded as an ideal proportion only during the 19th century, at which point the Romanticists projected it back to the medieval times that they so admired.

Is it true, then, that symmetry is an ideal state? Most of the conference participants were of the opinion that complete symmetry is rather boring. Only when a painting, or a piece of music or even a ballet, breaks its symmetries does art truly become interesting. A saying in Zen-Buddhism claims that real beauty appears only when symmetry is intentionally broken. The same holds for the sciences. Many phenomena arise at the borderline between symmetry and asymmetry. Pierre Curie, the noted French physicist and Nobel laureate, once said: *"C'est la dyssymétrie qui crée le phénomène"* ("It is dissymmetry that creates the phenomenon"). In the mid-19th century Louis Pasteur discovered that many chemical substances had "chirality." This term refers to the fact that right-handed and left-handed molecules may exist—mirror images of each other—which could not, however, be exchanged with one another, much in the same way that a right hand does not fit into a left glove.

A very sad reminder of the fact that left and right are not always interchangeable occurred during the 1960s. It was discovered that two versions of the pharmacological ingredient thalidomide, used in a medication called Contergan, existed—a right-handed version and a left-handed version. One form of the component is an efficient drug against nausea; the other causes the most dreadful birth defects.

The most significant symmetry break, according to one of the conference participants, occurred around 10 billion to 20 billion years ago. Matter and antimatter had been in equilibrium until, all of a sudden, something—nobody knows exactly what—happened that caused a disturbance in the symmetry. The result, so he claimed, was the Big Bang.

32

Random and Not So Random

In order to decide which team should kick off, a referee at a soccer game usually tosses a coin into the air and checks if it lands heads or tails. At a casino poker table, a croupier rolls a die and checks how many dots are embossed on the top side when it stops rolling. In the game of Lotto, a blast of air spews a bunch of Ping-Pong balls with numbers inscribed on them high into the air. At some point in time the machine swallows one of the balls and its number is recorded. In each of these cases, one may safely assume that the result of the throw was arrived at by pure chance. One can never predict the face of the coin, the dots on the die, or the number on the Ping-Pong ball.

Purists may point out that one side of a die or a coin might be heavier than the other and that this ever so slight weight difference may distort the results. But apart from this minor flaw, these objects generate fairly acceptable strings of random numbers—between 0 and 1 for coins, between 1 and 6 for dice, and between 1 and 45 for lottery Ping-Pong balls.

Random numbers are important not only in games or sports. There are several other areas in which these numbers are indispensable tools of the trade. In cryptography, for example, random numbers (actually randomly chosen prime numbers) are used to encrypt data. In engineering or economics random numbers are used for simulations. Instead of elaborately calculating a city's transportation flow by using probability theory, traffic situations can be tested with the help of simulations. A computer program could be written where a particular traffic light turns green whenever a randomly selected number is between 16 and 32, a truck approaches from the left, if the random number is odd, and so on. The simulation is run several thousand times, and an operator notes the observations: Do accidents occur? Are traffic jams caused? Since random numbers are reminiscent of the roulette game, such methods became known as Monte Carlo simulations. Even mathematics, the most exact science of all, can benefit from simulations. The volumes of complicated shapes, for example, can be determined by calling on Monte Carlo methods.

There is a problem, though, when generating random numbers with the methods described above. Techniques that involve throwing coins, dice, Ping-Pong balls, and other objects into the air are extremely inefficient. At best, about one number can be generated per second. To run high-quality simulations, millions and sometimes even billions of random numbers are required. It stands to reason that one would use computers to generate random numbers. After all they are able to produce huge amounts of numbers in just fractions of a second. But there is a snag. One of the great advantages of electronic computers—the ability to mindlessly execute prescribed instructions over and over again—turns out to be a devastating handicap when it comes to generating strings of random numbers. Starting with any number, computers will always calculate the next number, based on the preceding value. This means that in a sequence of computer-generated random numbers patterns will appear, and every number can, in principle, be predicted. The formula used to generate the "random" numbers may be complex and the pattern may be complex, but a pattern there is nevertheless. For good reason, computer-generated random numbers are called pseudorandom numbers. Even though they may pass a battery of tests for randomness, they are not truly random.

The trick that is used in the creation of pseudorandom numbers by computer involves using a random starting value, or seed. Once a seed has been selected, the program proceeds in a deterministic but unfathomable (to the user) way. It may calculate something along the lines of "take the third root of the preceding number, divide the result by 163, and then pick the 7th, 12th, and 20th digits after the decimal point." With this three-digit pseudorandom number in hand, the subsequent pseudorandom number can be calculated, and so on. Of course, since this example involves only three digits, no more than 1,000 different pseudorandom numbers can be generated. And whenever the computer encounters a three-digit number that has been used before, the program will proceed from there with an identical string of numbers. Hence cycles are invariably produced. Their onset can be delayed by making the pseudorandom numbers 15, 20, or more digits long, but in the end even the longest pseudorandom number sequence will end up in a cycle.

Whatever the size of the pseudorandom number, it is imperative that the signal which starts the process come from outside the computer. Otherwise the procedure would always start off with the same seed and all sequences generated by this program would be identical. Many things may serve as a starting signal: the time when the computer operator hits the "Enter" button on the keyboard; the operator's imperceptible, hence random, hand movement when he or she moves the computer mouse; and so forth.

But however carefully thought out the process might be, in the end all computer-generated random number sequences are of the "pseudo" kind. Scientists nevertheless thought that they could obtain satisfactory results and used their random number generators without many questions. In 1992, however, three physicists found to their horror that their simulations produced incorrect predictions, and hence the conclusions derived from their work were erroneous. Things got even worse. In 2003, two German physicists, Heiko Bauke and Stephan Mertens, proved that generators of random binary numbers produced too many zeros and not enough ones, due to the special role played by zero in algebra.

Organizations specializing in random numbers saw an opportunity. They decided to generate not only the starting value but all numbers outside the computer. The resulting strings of random numbers are put at the disposal of interested parties via the World Wide Web. The sources for these random numbers are natural phenomena such as the thermal crackling of transistors, the decay of radioactive material, the blubbering of lava lamps, or atmospheric background noise—all of which are completely, undeniably, incontestably random. These phenomena can be measured and registered with the help of Geiger counters, thermometers, or megaphones. Hence, genuine random numbers are being generated and not their "pseudo" versions.

33

How Can One Be Sure It's Prime?

Prime numbers are prized commodities when it comes to cryptographic schemes to keep digital messages, such as credit card numbers, secret on the Internet. The basis of most encryption methods is two very large primes that are multiplied together. The key to breaking the encrypted message is to retrieve the two factors of the resulting product—an undertaking that is not feasible because it would be enormously time consuming. Since even the fastest number-crunching computers would require days, weeks, or years to find the prime factors of a number that is a few hundred digits long, users of Internet business sites can rest assured that their credit card numbers are not accessible to anyone, provided of course that the right software is used. Only those who actually possess the key, that is, who know the prime factors, can decipher the encrypted message.

To use this encryption method, one needs to be sure that the numbers used for encoding are indeed primes. If they were not, they would be divisible into further components, and the factorization of the product would not yield a unique result. (Multiplying the two nonprimes 6 and 14 gives 84. This product could then be split into different pairs of factors—for example, 3 and 28 or 7 and 12.) In these cases some of the keys would be the correct ones, and some would be the wrong ones. So to avoid confusion, potential keys need to be certified as being prime before being used.

But proving that a number is prime is no easy feat. Existing algorithms that can certify that numbers are indeed prime are either very time consuming or, if they are fast, are able to certify the primality of an integer only with a certain probability. Practitioners yearn for an algorithm that is able to tell with certainty and in a speedy way whether a number is prime or not.

This is exactly what a team of three Indian computer scientists managed to devise. Manindra Agarwal and two of his students, Neerja Kayal and Nitin Saxena, at the Indian Institute of Technology in Kanpur utilized and extended Fermat's theorem—the so-called little theorem, not the more famous "last" theorem—to check for the primality of numbers. After devising their scheme, analysis of the computer program showed a most surprising result. The time required to check for primality did not grow exponentially with the number's size but merely had a polynomial run time.

It did not take more than a few days after the mathematicians announced their research results on the Internet until news agencies and media around the world picked up on the news. They celebrated the discovery as a breakthrough. The announcements were slightly exaggerated, even though on a theoretical level the three computer scientists did achieve a breakthrough. In mathematics the term "merely" (as in "merely" polynomial) is very relative. The running time of the algorithm that the Indian professor and his students proposed is indeed polynomial in N, with N representing the number of digits in the target integer. But it is proportional to N^{12}, which means that checking a 30-digit prime number—which in cryptographic terms is an extremely small key—requires on the order of 30^{12} computational steps.

Considering that the 1,000 largest primes known today all are over 40,000 digits long—the current world record lists a prime with 4 million digits—one quickly realizes that putting the discovered algorithm into practice is another matter altogether.

Nevertheless, this unexpected result created no small sensation among colleagues in the field. The three mathematicians had hit on a beautiful and fundamentally new idea. For practical purposes the algorithm is, admittedly, still too time consuming. But now that the ice has been broken, experts are confident that more efficient ways of calculation are imminent. This cautionary note aside, nobody need worry about whether a credit card number is secure. The discovered method cannot be applied to breaking cryptographic codes.

VI

Interdisciplinary Potpourri

34

A Mathematician Judges the Judges (Law)

Decisions handed down by the nine judges of the U.S. Supreme Court give rise to legal and political interpretations. Studies suggest that the justices are heavily influenced by political viewpoints, be they left-wing, right-wing, conservative, or liberal. But as Lawrence Sirovich, a mathematician at Mount Sinai School of Medicine in New York City, showed, it is possible to make a completely dispassionate, objective mathematical analysis of the verdicts. In an article published in the *Proceedings of the National Academy of Sciences,* he examined close to 500 rulings made by the Supreme Court presided by Chief Justice William Rehnquist between the years 1994 and 2002.

In principle, one can imagine two different kinds of courts and all variations in between. At one extreme, there is a bench made up of omniscient judges, who—were they to exist—know the absolute truth and therefore render their decisions completely unanimously. This court would be mathematically equivalent to a panel of judges who are completely influenced by economic and political considerations. On the assumption that they are influenced in the same manner, they too would cast identical votes. In both of these cases it would suffice to have one single judge, since the eight colleagues would simply be clones and therefore superfluous.

In contrast to this efficient but boring scenario, there is the situation, at the other end, where judges comply with the platonic ideal of making decisions completely independently of one another. They do not let themselves be swayed by political pressures, lobbyists, or their own colleagues. In this type of court none of the judges would be replaceable.

Of course, variations in between these two extremes are also possible. To find out which scenario exists in reality, Sirovich analyzed the so-called entropy of the judges' decisions. Entropy is a term from thermodynamics, which the mathematician Claude Shannon applied to information theory in the 1940s, that refers to the amount of "disorder" that exists in a system. When molecules are set within a crystal lattice—say, in ice—order is large and entropy is small. One would never be surprised to encounter a molecule at a certain point on the lattice. On the other hand, the random motion of molecules in a gas corresponds to disorder and hence to more entropy.

In information theory, entropy refers to the amount of information that exists in a signal. Applying the term to the decisions of judges implies the following: If all judges make exactly the same decision, order is at a maximum and entropy at a minimum. If, on the other hand, the judges make independent, random decisions, entropy is large. For this reason entropy can also be used as a measure of the information content of the decisions: If judges make identical decisions, the information content is small. It would suffice to obtain the ruling of a single judge, and the opinions of the other judges would be superfluous.

Sirovich calculated the information content of the 500 rulings of the Rehnquist court by computing the correlation between the decisions handed down by the different judges. It turned out that the rulings were far removed from a random distribution. Close to half of the decisions were unanimous, not necessarily because the judges were being influenced by any outside forces but rather because many cases turned out to be straightforward from a legal point of view. Added to this, Justices Scalia and Thomas voted identically in over 93 percent of the cases that reached them. Whenever one of these two justices reached a decision, it was more than likely that the other justice would make the same decision. It was also common knowledge that in many cases Justice Stevens's decisions deviated from the majority, and this too gradually ceased to surprise anyone.

Sirovich's work showed that, on average, 4.68 judges (yes, we are speaking fractional judges here) made their decisions independently of one another. In other words, they act as "ideal" adjudicators. This implies that the other 4.32 judges are actually superfluous, since their rulings were, in general, determined by the other judges. The figure of 4.68 is encouraging, says Sirovich, since it proves that the majority of the judges are in large measure independent of one another. They are not influenced by particular worldviews nor are they pressured by each other to make a unanimous decision. While one may believe that the ideal number of independent judges would be nine, this is, in fact, not true. Nine independent judges would imply that each decision is handed down randomly. This would only be possible, however, if the judges completely ignored the facts before them and the legal arguments. Obviously it would not be in the interest of justice or in keeping with the spirit of the law to have judges who could be replaced by random number generators.

Then Sirovich moved on to a different mathematical calculation. The question he now asked was how many judges would have been necessary to hand down 80 percent of the verdicts in the same manner that the Supreme Court of nine judges had done. Each ruling of a nine-justice court (yes, no, yes, yes, no, . . .) can be considered a point in nine-dimensional space. But since the rulings of certain judges correlate to some degree, this space can be reduced. To see how much the space can be reduced, Sirovich utilized "singular value decomposition" from linear algebra. This is a technique that has been successfully applied to such diverse subjects as pattern recognition and analyses of brain structure, chaotic phenomena, and turbulent fluid flow. Sirovich's analysis showed that 80 percent of the rulings could be represented in a two-dimensional space. Hypothetically, then, only two virtual judges would have been required—albeit composed of different parts of the nine existing judges—in order to hand down four-fifths of all rulings.

35

Elections Aren't Decided by the Voters Alone (Political Science)

Switzerland is known as one of the world's foremost democracies. In fact, when the United States sought a framework of government for its 13 states in the late 18th century, it was the Swiss model of cantons that was adopted. Switzerland consists of 25 cantons, each one of which wants, and gets, a say in the affairs of state. Every 10 years the citizens of all the cantons elect their delegates to the Federal Council. Article 149 of the Swiss Federal Constitution stipulates among other things that the Federal Council consists of 200 representatives and that the seats are to be allocated in proportion to the population of each canton.

One may think that nothing could be easier than to fulfill these clear stipulations, but this would be far from the truth. The constitution's apodictic instructions cannot, in general, be obeyed. The reason is that each canton can only send an integer number of representatives to the Federal Council. Let us take the canton of Zurich, for example. At last count it had a population of 1,247,906, which corresponded to 17.12 percent of the Swiss population. How many representatives can Zurich delegate to the Federal Council? For a house size of 200, can the canton send 34 or 35 representatives to the capital, Berne? And once the issue of regional representation has been resolved, how should the 34 or 35 seats be apportioned to the parties that participated in the elections in Zurich?

One simple way to allocate seats in parliament is by rounding the results. This method is unsatisfactory, however, since rounding up or down too often can lead to a change in the total number of seats, thus violating article 149 of the constitution. So some other method must be found.

According to theoreticians who deal with apportionment problems, a fair method of allocating seats in a council must fulfill two requirements. First, the method must yield numbers of seats that equal the computed numbers, rounded either up or down. Hence, Zurich would be apportioned not less than 34 and not more than 35 representatives. This is the so-called quota rule. Second, the allocation method must not produce paradoxical results. For example, the number of seats allocated to a growing constituency must not decrease in favor of a constituency whose population decreases. This is called the "monotony requirement."

At first glance, the requirements appear to be reasonable. But when they are investigated mathematically or tested in practice, they are nothing of the sort. In 1980 the mathematician Michel Balinsky and the political scientist Peyton Young proved a very sobering, if disappointing, result: No ideal method of apportionment exists. A method that satisfies the quota rule cannot satisfy the monotony requirement, and any method that satisfies the monotony requirement fails to satisfy the quota rule.

So what is to be done? Article 17 of Swiss federal law spells out how the seats of the Federal Council are to be apportioned to the cantons. First, the cantons whose populations are too small to warrant a delegate of their own are apportioned one seat each. Then the number of seats that the other cantons are due is computed, fractions and all. Next, a preliminary distribution takes place: Each canton receives the rounded-down number of seats. Finally, in the so-called remainder distribution, the leftover seats are allocated to those cantons whose dropped fractions are the largest. The method seems acceptable even though it favors larger parties ever so slightly: Rounding 3.3 seats down to 3 is more painful than rounding 28.3 down to 28.

But even though the method seems so sensible, it can cause enormous problems. This first came to light in the United States, which used the same allocation method in the 1880s. A conscientious clerk who also happened to be good with numbers realized, to his utter amazement, that

TABLE 1 The Alabama Paradox (State A loses a seat, even though the size of the Federal Council is increased.)

	State A	State B	State C	Total
Council with 24 seats				
Population	390	700	2,700	3,790
Proportion	10.29%	18.47%	71.24%	
Proportion x seats	2.47	4.43	17.10	
Preliminary distribution	2	4	17	23
Remainder	0.47	0.43	0.10	
Remainder distribution	1	0	0	
Total number of seats	3	4	17	24
Council with 25 seats				
Population	390	700	2,700	3,790
Proportion	10.29%	18.47%	71.24%	
Proportion x seats	2.57	4.62	17.81	
Preliminary distribution	2	4	17	23
Remainder	0.57	0.62	0.81	
Remainder distribution	0	1	1	
Total number of seats	2	5	18	25

the state of Alabama would lose a representative if the size of Congress were increased from 299 seats to 300. This nonsensical situation, which contradicted the monotonicity condition, was henceforth called the Alabama paradox. In the example illustrated in Table 1 an increase in the size of the House of Representatives from 24 to 25 has the ridiculous consequence that State A loses a representative.

As far as Switzerland is concerned, this problem ceased to exist in 1963 when the number of representatives, which heretofore had varied, was fixed at 200. In the United States, the size of Congress has been fixed at 435 representatives since 1913.

The Alabama paradox is not the only potential problem, though. Another paradox, referred to as the population paradox, may appear under certain circumstances. Constituencies whose populations have increased could lose representatives in favor of other constituencies whose populations have decreased. In the example in Table 2, State C, whose population decreased, is awarded an additional representative at the expense of State A, whose population increased. In this paradox and in the Alabama paradox, the culprits are the fractional parts of seats. The population paradox still looms in Switzerland, albeit without having caused any problems so far. And in the spirit of "if it ain't broke, don't fix it," the issue has been put aside. For the time being, the method of rounding down and then distributing the remaining seats according to the largest fractions is still in force.

But the problems are not over, even after the 200 seats of the Federal Council have been apportioned to the cantons. Each canton's seats must now be assigned to the individual parties. Articles 40 and 41 of Swiss federal law say how this is to be done. The key for the distribution is based on the proposal of Victor d'Hondt (1841–1901), a

TABLE 2 The Population Paradox (The population of State C decreases. Nevertheless it gains a seat at the cost of the growing State A.)

	State A	State B	State C	Total
Parliament with 100 seats				
Census 1990	6,570	2,370	1,060	10,000
Proportion	65.7%	23.7%	10.6%	
Preliminary distribution	65	23	10	98
Remainder	0.7	0.7	0.6	
Remainder distribution	1	1	0	2
Total number of seats	66	24	10	100
Census 2000	6,600	2,451	1,049	1,0100
Proportion	65.35%	24.26%	10.39%	
Preliminary distribution	65	24	10	99
Remainder	0.35	0.26	0.39	
Remainder distribution	0	0	1	1
Total number of seats	65	24	11	100

Belgian lawyer, tax expert, and professor for civil rights and tax law at the University of Gent.

D'Hondt suggested a rule which assures that the greatest number of voters stand behind each seat. It works as follows: For each seat the number of votes cast for a party is divided by the number of seats already allotted to the party, plus one. The seat then goes to the highest bidder. The process continues until all seats have been filled. (Table 3 should make this complicated-sounding procedure clearer.)

It did not take long for the Swiss to realize they had reinvented the wheel. As it turned out, d'Hondt's method is computationally equivalent to the method that had been put forth a hundred years earlier by President Thomas Jefferson. He had used the method to apportion delegates to the House of Representatives in the United States. As far as the Swiss were concerned, they refused to relin-

TABLE 3 Jefferson–d'Hondt–Hagenbach-Bischoff Method (Ten seats are to be allocated.)

	List A	List B	List C
Votes	6,570	2,370	1,060
1. Seat	6,570*	2,370	1,060
2. Seat	3,285*	2,370	1,060
3. Seat	2,190	2,370*	1,060
4. Seat	2,190*	1,185	1,060
5. Seat	1,642*	1,185	1,060
6. Seat	1,314*	1,185	1,060
7. Seat	1,095	1,185*	1,060
8. Seat	1,095*	790	1,060
9. Seat	938	790	1,060*
10. Seat	938*	790	530
Total	7	2	1

NOTE: The number of votes of each list is divided by the number of seats it already has been allocated, plus one. The highest bidder (denoted by *) receives the seat, until all seats have been allocated.

quish ownership of the title to either the Belgians or the Americans and decided to name the method after Eduard Hagenbach-Bischoff, a professor of mathematics and physics at the University of Basle. Hagenbach-Bischoff had came across the method while serving as councillor of the city of Basle.

The fact that the Jefferson–d'Hondt–Hagenbach-Bischoff method slightly favors larger parties was not considered a major flaw. In fact, only when it is applied in a cumulative fashion will a disadvantageous impact be felt by smaller parties—for example, when a house uses the d'Hondt method a second time to fill various commissions. Paraphrasing Winston Churchill, one could say that the Jefferson–d'Hondt–Hagenbach-Bischoff method is the worst possible form of allocating seats . . . except for all the others that have been tried.

36

A Dollar Isn't Always Worth a Dollar (Insurance)

In 1713 Nikolaus Bernoulli (1687–1759), the famous Swiss mathematician, posed a question about the following game:

— Toss a coin.
— If it shows heads, you get \$2 and the game is over. Otherwise you toss again.
— If the coin now shows heads, you get \$4, and so on. Whenever you toss tails the prize is doubled.
— After *n* tosses the player gets $\$2^n$ if heads appears for the first time.

After 30 tosses this is a sum of more than \$1 billion—a gigantic prize. Now the question: How much would a gambler pay for the right to play this game? Most people would offer between \$5 and \$20, but is that reasonable? On the one hand, the chance of winning more than \$4 is just 25 percent. On the other hand, the prize could be enormous because the probability of tossing a very long series of tails before that first toss of heads, although very small, is by no means zero. The huge prize that could be won in this case compensates for the very small probability of success. Nikolaus Bernoulli found that the expected prize is infinite! (The expected prize is calculated by multiplying all the possible prizes by the probability that they are obtained and adding the resulting numbers.)

And therein lies the paradox: If the expected prize is infinite, why is nobody willing to pay \$100,000, \$10,000, or even \$1,000 to play the game?

The explanation of this mysterious behavior touches the areas of statistics, psychology, and economics. Two other Swiss mathematicians, Gabriel Cramer (1704–1742) and Nikolaus's cousin Daniel Bernoulli (1700–1782), suggested a solution. They postulated that $1 does not always carry the same "utility" for its owner. One dollar brings more utility to a beggar than to a millionaire. Whereas to the former, owning $1 can mean the difference between going to bed hungry at night or not, the latter would hardly notice the increase of his fortune by $1. Similarly, the second billion that one would win if the 31st toss showed tails wouldn't carry the same utility as the first billion that one would receive already after 30 tosses. The utility of $2 billion just isn't twice the utility of $1 billion.

Herein lies the explanation of the mystery. The crucial factor is the *expected utility* of the game (the utilities of the prizes multiplied by their probabilities), which is far less than the *expected prize*. Daniel Bernoulli's treatise was published in the *Commentaries of the Imperial Academy of Science of St. Petersburg,* and this surprising insight was henceforth called the St. Petersburg paradox.

Around 1940 the idea of the utility function was taken up by two immigrants from Europe working at the Institute for Advanced Study in Princeton, New Jersey. John von Neumann (1903–1957), one of the outstanding mathematicians of the 20th century, was Jewish and had been forced to flee his native Hungary when the Nazis invaded the country. The economist Oskar Morgenstern (1902–1976) had left Austria because he loathed the National Socialists.

In Princeton the two immigrants worked together on what they thought would be a short paper on the theory of games. But the treatise kept growing. When it finally appeared in 1944 under the title *Theory of Games and Economic Behavior,* it had attained a length of 600 pages. This pioneering work was to have a profound influence on the further development of economics. In the book Bernoulli and Cramer's utility function served as an axiom to describe the behavior of the proverbial "economic man." However, it was soon noticed that in situations with very low probabilities and very high amounts of money, test candidates often made decisions that contravened the postulated axiom. The economists remained unfazed. They insisted that the theory was correct and that many people were simply acting irrationally.

Despite its shortcomings, utility theory has had a far-reaching consequence. The explanation that Bernoulli and Cramer offered for the St. Petersburg paradox formed the theoretical basis of the insurance business. The existence of a utility function means that most people prefer having $98 in cash to gambling in a lottery where they could win $70 or $130, each with a chance of 50 percent—even though the lottery has the higher expected prize of $100. The difference of $2 is the premium most of us would be willing to pay for insurance against the uncertainty. That many people buy insurance to avoid risk yet at the same time spend money on lottery tickets in order to take risks is another paradox, one that is still awaiting an explanation.

37

Compressing the Divine Comedy (Lingusitics)

The amounts of data we want to store on our hard disks expand even faster than the rapidly increasing capacity of the storage media. Therefore, software solutions are sought that enable us to pack data ever more densely onto the disks, thereby overcoming hardware limitations. But compression techniques can have unexpected applications.

To understand data compression, it is necessary to understand the notion of entropy. In physics, entropy is a measure of the disorder in a system—for example, in a gas. In telecommunications, entropy is a measure of the information content of a message. A message consisting, for example, of 1,000 repetitions of the number 0 has very little information content and a very low entropy. It can be compressed to the concise formula "1,000 times 0." On the other hand, a completely random sequence of ones and zeros has very high entropy. It cannot be compressed at all, and the only way to store this string is to repeat every character.

Relative entropy indicates how much storage space is wasted if a sequence of characters is compressed with a method that was optimized for a different sequence. The Morse code, optimized for the English language, may serve as an example. The letter occurring most frequently in English, "e," was allocated the shortest code: a dot. Letters that appear rarely are allocated longer codes—for example, "– – . –" for "q." For languages other than English the Morse code is not ideal because the lengths of the codes do not correspond to the frequency of the letters. Relative entropy then measures how many additional dots and dashes are needed to transmit, say, an Italian text with a code that is optimal for the English language.

Most data compression routines are based on an algorithm that was developed in the late 1970s by two Israeli scientists from the Technion in Haifa. The method that Abraham Lempel, a computer scientist, and Jacob Ziv, an electrical engineer, devised relies on the fact that very often identical strings of bits and bytes reoccur in a file. At a string's first appearance in the text, it is entered into a kind of dictionary. When the same string reoccurs, a pointer simply refers to the appropriate place in the dictionary. Since the pointer takes up less space than the sequence itself, the text is compressed. But there is more. Preparation of the table that lists all the strings does not follow the compilation rules of a standard dictionary. Rather it adapts itself to the specific file that is to be compressed. The algorithm "learns" which strings occur most often and then adapts the compression to it. With increasing file size, the space needed to store it decreases toward the text's entropy.

There are no limits to the imaginative use of computers in science. Compression algorithms too can be applied to areas other than the space-saving storage of computer files. Two mathematicians and a physicist—Dario Benedetto, Emanuele Caglioti, and Vittorio Loreto from the University La Sapienza in Rome—decided to put the Lempel–Ziv algorithm to work. Their aim was to identify the authors of pieces of literature. Ninety texts written by 11 Italian authors (including Dante Alighieri and Pirandello) served as the raw material. The text of a certain author was chosen, and two short texts of equal lengths were appended: one from the same author and one from another author. Both files were fed into compression programs—such as the ubiquitous WinZip program—and the scientists checked how much storage space each of them required. They conjectured that the relative entropy of the combined text would give an indication as to the authorship of the anonymous text. If both texts were written by the same author, the algorithm would require less storage space than if the attached text was written by a different author. In the latter case the relative entropy would be higher since the algorithm would have to consider the different styles and different words used by the two authors. As a consequence, it would use more space to store the file. Hence, the shorter the compressed file of the combined texts, the more likely it was that the original text and the appended text derived from the same author.

The results of the experiment were nothing less than astounding. In close to 95 percent of the cases the compression programs allowed the correct identification of the author.

In their excitement over the success of their newly found approach the three scientists failed to notice, or at least forgot to mention in their bibliography, that their method was not quite as novel as they had imagined. In fact, they were not the first ones to think that mathematical methods could be used to attribute literary texts to their authors. George Zipf, a professor of linguistics at Harvard, had already studied such issues as word frequency in 1932. And the Scotsman George Yule had shown in 1944, in a paper entitled "The Statistical Study of Literary Vocabulary," how he was able to attribute the manuscript "*De imitatione Christi*" to the well-known mystic Thomas à Kempis, who lived in Holland in the 15th century. And of course mention must be made of the "Federalist Papers" from the 18th century, whose authorship (Alexander Hamilton, James Madison, and John Jay) was ascertained in 1964 by American statisticians R. Frederick Mosteller and David L. Wallace.

Since everything had worked out extremely nicely, Benedetto, Caglioti, and Loreto decided to undertake another experiment. They analyzed the degrees of affinity between different languages. Two languages that belong to the same linguistic family should have low relative entropy. Therefore it should be possible to compress a combination of two texts more efficiently if they are written in languages that are related than if they are written in two languages that do not belong to the same family. The scientists analyzed 52 different European languages. Again their efforts were successful. Using the zipping program, they were able to attribute each tongue to its correct linguistic group. Italian and French, for example, have low relative entropy and therefore belong to the same linguistic group. Swedish and Croatian, on the other hand, have high relative entropy and must come from different linguistic groups. WinZip was even able to identify Maltese, Basque, and Hungarian as isolated languages, which do not belong to any of the known linguistic groups.

The success of their experiments made the three scientists optimistic that measuring relative entropy with zipping software may also work with other data strings, like DNA sequences or stock market movements.

THE PROOF OF THE PUDDING

The method described above inspired this writer to a test. The text samples were articles I wrote as part of my journalistic brief for the *Neue Zürcher Zeitung*, a major Swiss daily newspaper. There were 18 pieces covering events in Israel and comprising some 14,000 words and 105,000 characters. After removing all titles and subtitles, the texts were stored as an Ascii file and compressed by WinZip. I took one look at the result and had the fright of my life. The compression had downsized the original texts, which I had written with painstaking effort in the course of an entire month, by a full two-thirds. The inescapable conclusion was that only 33 percent of the original texts contained significant information, while two-thirds were pure entropy. In other words, they were completely redundant.

I sought to console myself with the argument that it must be the competent sequencing of his words that provided significant information and not the text as such. To prove this face-saving thesis I put the 14,000 words into alphabetical order and then compressed them again. Lo and behold: The alphabetically ordered sequence of words could be compressed by more than 80 percent and thus offered only 20 percent information. (This, of course, was not surprising, since the words "Israel" or "Israeli" appeared some 231 times, and the word "Palestinian" in all its derivations appeared a total of 195 times.) What this shows is that putting the words into sensible order—in a way that could only be done by a competent journalist—offers some 13 percent more information than a simple dictionary. It was not a huge consolation, but it sufficed to let me breathe a sigh of relief.

But another blow was to come. A randomized collection of the 14,000 words could only be compressed by 60 percent. Comparing that to the compression rate of 66 percent for the actual texts left the distinct impression that a totally random collection of words contained more information than did the actual articles.

For the actual experiment three sample texts were used:

two long articles containing 1,000 words each, written by me and by Stefan, an editor of the newspaper, and a short one, containing 50 words, written by me. The shorter text was appended to each of the two longer texts and both files were compressed.

The results matched those gathered by the Italian scientists: When my short text, which comprised 462 characters, was attached to my long text, WinZip required 159 additional characters. When attached to Stefan's long text, the compression program needed another 209 characters. Thus it was proven that the shorter text was written not by Stefan but by yours truly.

38

Nature's Fundamental Formula (Botany)

D'Arcy Thompson (1860–1948) was a Scottish biologist, mathematician, and classics scholar well known for his varied interests and mildly eccentric habits. Nowadays he is probably best remembered for his pioneering work, *On Growth and Form*, published in 1917, in which he demonstrates that mathematical formulas and procedures can describe the forms of many living organisms and flowers. Mussels, for example, can be visualized as logarithmic spirals, and honey webs can be understood as the shapes with the smallest circumference from among all rectangles that cover an area without gaps.

But d'Arcy Thompson's most astounding observation must have been that the shapes of quite different looking animals are often mathematically identical. Using the right coordinate transformation—that is, by tugging, pulling, or turning—a carp can be transformed into a moonfish. This also goes for other animals. The profiles of many four-legged creatures and birds differ from one another only because of the varying lengths and angles in their shapes.

D'Arcy Thompson's explanation for this phenomenon was that different forces pull and squeeze the body until it assumes a streamlined figure or some other shape that is suitable for its environment. "Everything is the way it is because it got that way," he wrote. The transformed shape is then passed on to the next generation. Since this can be interpreted as an adaptation to the environment, d'Arcy's observations fit seamlessly into the already popular Darwinist worldview. (As an aside, the technique of emphasizing and distorting facial or bodily features—ears that stick out, oblong-shaped heads, big noses—has been the caricaturist's tool of the trade for centuries, to readers' immeasurable delectation.)

The learned Scotsman was not the first to have used mathematics to describe natural phenomena. At the turn of the 13th century, Leonardo Bonacci from Pisa, later known as Fibonacci (son of Bonacci), had studied how rabbits multiply. The question he asked was how many pairs of rabbits coexist at any point in time, beginning with one pair of baby rabbits. At the end of the first month, the pair has reached puberty and the two rabbits mate. There is still only one pair of rabbits, but at end of the second month, the female gives birth to a new pair. There is now one adult pair and one baby pair. The adults mate again, thus producing another pair at the end of the third month. At this point there are three pairs of rabbits. One of these pairs has just been born, but the other two are old enough to mate, which, being rabbits, they do. A month later both they and their parents produce one additional pair each, and now a total of five pairs coexist. The answer to Bonacci's question turns out to be a number sequence that from then on was called the Fibonacci series. Here are the first entries of the series: 1, 1, 2, 3, 5, 8, 13, 21, 34. . . . The process continues indefinitely, each number in the series being computed as the sum of the two predecessors (for example, 13 + 21 = 43). Of course, this does not prove that rabbits will take over the world, but only that Fibonacci had forgotten to factor in that rabbits tend to die after a certain time. (What this example does prove is that despite faultless induction mathematicians can sometimes abuse their science and arrive at erroneous conclusions simply by starting off with an incorrect premise.)

The Fibonacci series turns up in many guises. Kernels of sunflowers, for example, are arranged in left- and right-turning spirals. The number of kernels in the spirals usually corresponds to two consecutive numbers in the Fibonacci series—for example, 21 and 34. The number of spirals on pines and pineapples or the number of pricks on a cactus also corresponds to two consecutive numbers in the Fibonacci series. Nobody quite knows why this is so, but there is a suspicion that the phenomenon is somehow connected to the efficacy of the plants' growth.

Johan Gielis, a Belgian botanist who in the past was preoccupied with bamboo trees, decided to join the long line of scientists whose ambition it is to reduce natural phenomena to a single simple principle. His article, published in *The American Journal of Botany*, quickly caught the interest of the public, due in no small measure to the activation of a strong publicity machine and the catchy phrase—"the Superformula"—used in the article. In his work Gielis purports to show that many shapes found in living organisms can be reduced to a single geometrical form.

He starts off with an equation for a circle which, by adjusting a number of parameters, can be transformed into the equation for ellipses. Additional variations of the equation generate other shapes—triangles, squares, star shapes, concave and convex forms, and many other figures. Instead of tugging and tweaking at pictures as d'Arcy Thompson had done many years previously, Gielis tugged and tweaked at the six variables of his Superformula, thus simulating pictures of different plants and organisms. Because all the shapes that emerge do so through the transformations of the circle, they are, Gielis maintains, identical to each other.

The Superformula can by no means be defined as higher mathematics. Nor does it offer revolutionary insights or discoveries. Despite the media fuss and the accolades bestowed on it, the Superformula belongs more to hobby mathematics than to anything serious scientists could get excited about. At least Fibonacci had tried to give an explanation for his series in terms of the procreation of rabbits. And d'Arcy Thompson had come about his transformations by studying forces that supposedly act on an organism's body. Gielis's Superformula, on the other hand, offers no explanation whatsoever. It merely gives approximate descriptions of a number of organic forms. This drawback, however, did not keep the author from patenting the algorithm for his formula and creating a company to develop and market his invention.

39

Stacking Words Like Oranges and Tomatoes (Computer Science)

In August 2002 the International Congress of Mathematicians was held in Bejing. This mammoth convention, which takes place every four years and to which thousands of mathematicians come from all corners of the world, is a good opportunity for bestowing prizes on the worthiest among the many worthy. Since 1982 a prize called the Nevanlinna Prize (named after the Finnish mathematician Rolf Nevanlinna) has been awarded for outstanding work in the field of theoretical computer science. In 2002 it was bestowed on Madhu Sudan, a professor at the Massachusetts Institute of Technology. The laudation mentioned, among other things, Sudan's fundamental contributions to the area of error-correcting codes.

Computer users are probably familiar with error-correcting software. Most word-processing programs contain this feature in the form of a spell checker. If, for example, you type "hte," a wavy red line under the senseless letter combination will indicate that such a word does not exist in the English language. Commonly this is referred to as error detection.

Some word-processing programs go further than that. In the English language, for instance, only one legitimate word exists containing only the three letters t, h, and e. Hence there is no doubt about which word the typist intended to type. Accordingly, an advanced word-processing program will automatically replace the incorrect sequence of characters with "the." In this case an error correction took place.

It is much simpler to detect errors than to correct them. When transmitting text or copying a file, it suffices to include a check sum to detect errors. For a string of numbers, for example, this quality check could be in the form of the sum of the string's digits. If the "check sum" at the points of transmission and reception do not correspond, one must conclude that at least one error has crept in, and the number sequence needs to be retransmitted.

This repetition of a transmission is time consuming. It certainly would be more efficient to correct the errors at the point of reception automatically. So, instead of having to retype the word "the," a good program automatically replaces the erroneous character sequence "hte" with the correct one.

How and when is this possible? Surprisingly this problem is closely connected to a totally different problem that Johannes Kepler, an astronomer and mathematician who lived in the first half of the 17th century, investigated at great length. The question he asked was, which is the most efficient way to arrange oranges or tomatoes on a greengrocer's cart—that is, with the minimum amount of gaps in between them?

How can the problem of stacking tomatoes have anything to do with error correction? Imagine a three-dimensional space with the letters of the alphabet arranged at equal distances, say 1 centimeter, from each other along the three axes. Three-letter words can be visualized as points in this space, the x axis representing the first letter, the y axis the second, and the z axis the third letter of the word. Thus each word is represented by a point within a cube measuring $26 \times 26 \times 26$ centimeters. Points that do not correspond to any word in the English language remain empty. Hence, if a nonsensical character sequence is transmitted, this would correspond to an empty space that an error detection program would automatically flag. An error correction program does more: It automatically seeks out the nearest "legitimate" word.

The more distant from each other the legal spots are, the less doubt exists regarding possible errors and the easier it is to replace a faulty sequence of characters with the correct one. Hence, to avoid ambiguities, it is important to ensure that no other proper word lies within a certain distance around every legitimate word. On the other hand, one wishes to cram as many words as possible into the cube. Hence, the problem that presents itself is as follows: How can as many points as possible be stored in the cube without the points lying closer to each other than a minimal distance? Thus we have arrived at the question that Kepler asked himself: How can oranges or tomatoes be stacked as densely as possible without squashing each other?

For centuries greengrocers were well aware of the fact that their fruits and vegetables are most efficiently packed in piles and layers that are arranged in a honeycomb or hexagonal pattern. It was only in 1998, however, that this conjecture was actually proved rigorously by the American mathematician Thomas Hales. In theory, 17,576 three-letter words can be stored along the lattice points of a cube holding 26 letters ($= 26^3$). The answer to Kepler's question shows, however, that the densest stacking of words, with 1 centimeter between any two of them, would allow about 25,000 words to be stored in the same space. If, on the other hand, one wanted to establish a broader security band, say 2 centimeters, between any two proper words, then the most efficient stacking method allows for only about 3,000 words.

To deal with words consisting of more than three letters, one simply turns to spaces of more than three dimensions. However, the most efficient method of packing has so far not been determined for all higher-dimensional spaces.

40
The Fractal Dimension of Israel's Security Barrier (Geography)

Politics should play no role in mathematics, but mathematics is ubiquitous, even in politics. Take the security fence that Israel is building on the West Bank. Its legality was examined by the International Court of Justice, but not only is the course of the construction in dispute, the two sides cannot even agree on a simple fact: its length.

An Israeli Army spokesman declared that the barrier around Jerusalem would be 54 kilometers long, but Khalil Toufakji, a geographer at the Center for Palestinan Studies in Jerusalem, said he had checked the Army's data and reached the conclusion that it would be 72 kilometers long.

For once both sides could be right—or wrong. The reason lies in the mathematical theory of fractals, which describes geometric patterns that are repeated at ever smaller scales. The length of the wall depends on the scale of the map used: Where 1 centimeter on paper corresponds to 4 kilometers in nature (scale 1:400,000), the barrier is only about 40 kilometers long. On a more detailed map, on which 1 centimeter corresponds to 500 meters (scale: 1:50,000), more details of the barrier's meandering course can be discerned, and its length jumps to 50 kilometers. On a scale of 1:10,000 even more details become visible and the wall seems to become longer still.

Now take a walk along the concrete blocks that make up the barrier. One soon notices that the wall often curves around single houses, snakes between two fields, or avoids topographical obstacles. Those details are too small to be included even on large-scale maps. Hence the physical length of the barrier is longer than can be indicated even on the most exact maps. All of a sudden, the differences in the declarations of the Israelis and the Palestinians start to add up.

It all started with an article by the French mathematician Benoit Mandelbrot. In his 1967 paper "How Long Is the Coast of Great Britain?," Mandelbrot did not even try to answer his question. Instead he established that it has no meaning. On large-scale maps of Great Britain, bays and inlets are visible that cannot be discerned on less detailed maps. And if one checks out the cliffs and beaches on foot, a longer coastline emerges, its exact length being determined by the water level at the moment of measurement.

The observation also holds for land-based frontiers. Except for geographical borders defined as straight lines, as for example between North and South Dakota, there is no "correct" length of a border. In Spanish and Portuguese textbooks, for example, the lengths of the common border deviate by up to 20 percent. This is because the smaller state uses larger-scale maps to depict the fatherland, which results in longer borders.

The only quantitative statement that can be made, according to Mandelbrot, is about the line's "fractal dimension." This is a number that, in a way, describes the jaggedness of a geometrical object. For all coastlines and borders the fractal dimension is between 1 and 2. The more a line winds and meanders, the higher its fractal dimension. The border between Utah and Nevada has the fractal dimension 1, as we expect from regular lines. The British coastline has a fractal dimension of 1.24, and the even more jagged Norwegian coast has a fractal dimension of 1.52.

Fractal theory applies not only to lines on surfaces but also to surfaces in space. If the alpine landscape of Switzerland were ironed flat, for example, this country could well be as large as the Gobi Desert. A few years ago two physicists computed that the surface of Switzerland has the dimension 2.43. With this value it falls about halfway between the flat desert, which has a dimension of 2.0, and three-dimensional space.

With his bewildering article Mandelbrot heralded the age of fractals. Soon the strange shapes were discovered everywhere in nature: in trees and ferns, blood vessels and bronchia, broccoli and cauliflower, lightning, clouds and snow crystals, even in the movements of stock markets.

As for the West Bank fence, maybe it's just as well it's convoluted. If it were completely straight, it could theoretically reach all the way from Beirut to Mecca. Then the International Court of Justice would really have something to ponder.

41

Calculated in Cold Ice (Physics)

Two physicists from the University of Hokkaido in Sapporo, Japan, investigated a phenomenon that—come Christmas time—many young children observe every year. Looking out of the window they may notice icicles hanging from the roof gutters. The more curious among them may wonder about the ringlike ripples around the icicle, which occur at regular distances along its shaft. The two physicists had outgrown their childhood years but kept their curiosity. They could not help themselves but sit down and try to work out how the phenomenon comes about. The first thing they found was that the distance between two peaks always measures about 1 centimeter, independent of the length of the icicle and the outside temperature. The two scientists then developed a theoretical model that explains the surprisingly universal structure of icicles.

An icicle grows as thin sheets of water flow down its shaft. Part of the water freezes. The rest drips from the icicle's tip. But the ice that is left behind does not build up uniformly. What the two physicists found was that two counteracting effects account for the mysterious ripple phenomenon.

The first effect is the so-called Laplace instability, which says that more ice builds up on the convex parts of the icicle's surfaces than on the concave parts. This happens because protruding parts of the icicle are more exposed to the weather, and therefore more heat is lost there than from the somewhat protected indentations. This is why the ripples gradually grow thicker at these particular spots of the icicle. Another effect prevents these buildups from growing indefinitely, however. The Gibbs–Thomson effect says that the thin water layer that flows down the icicle's shaft has a counterbalancing effect on the temperature, thus inhibiting a massive growth of the ripples.

With the help of no less than 114 equations the two researchers reached the conclusion that the peak-to-peak distance between the ripples must measure about 1 centimeter. Their analysis also yielded a prediction. The ripples should gradually migrate down the icicle at about half the speed that the icicle grows—a phenomenon the researchers hope will soon be verified experimentally.

The two Japanese physicists were not the first to be interested in hibernal phenomena. Four hundred years earlier the astronomer Johannes Kepler noticed that snowflakes always have hexagonal patterns, even though no two are ever alike. Pleased with his discovery, he set about writing a booklet that he presented the following New Year to a friend. In "A New Year's Gift or On the Six-Cornered Snowflake," Kepler attempted to explain the phenomenon. The learned scientist suggested several possible causes for the mysterious shape. At first he tried to find a relationship between the hexagonal shapes of ice crystals and honeycombs but was unsuccessful. Realizing that he did not have the means to answer his question, he finally gave up. In the closing words to his booklet he remarked that one day chemists would surely discover the true cause for the snowflakes' sixfold symmetry. Kepler was right, but it took another 300 years, until the beginning of the 20th century, when the German scientist Max von Laue invented X-ray crystallography. Only with this new tool could the secret of the flakes' crystal structure be made visible and explained.

Snow crystals form around dust particles that are swirled upward into the atmosphere. When they float back down in the humid air, water molecules attach themselves to the nucleus, that is, the dust particle. Like nearly everything in nature, the molecules attempt to attain the state of lowest energy. At temperatures between 12 and 16 degrees Celsius below freezing, this state is achieved when the water molecules are arranged in a lattice such that each molecule is surrounded by four others in a pyramid-like structure. Looking at the structure from above with the help of X-rays, it resembles a hexagon. The tips of the hexagon branch out ever so slightly into space, but that suffices for what follows: water molecules, whirling around in the air, love to land on one of these protrusions. The tips, or snowflake arms, then keep growing, like dendrites, until even the naked eye can make them out. This is how snowflakes form.

42

Built on Sand (Physics)

It is summer and many of us are drawn to wander down to the beach. Children happily unpack their buckets and spades, settle close to the water, and build little piles of sand. Gradually the mounds get bigger until—swoosh—avalanches shrink them to near nothing. Undiscouraged, the kids start over again, building more piles until they too are destroyed by avalanches. What a great opportunity for Dad to get his hands dirty and give one of his educational lectures. If only they would pay attention, he groans, surely the kids could succeed in building a higher sandcastle. But what does he know? No matter how careful, sand avalanches always make the piles fall apart. It does not help to allow the sand to run ever so slowly through one's fingers. Amazingly, avalanches always occur when the piles have reached a certain height.

What hides behind these jolly summer activities is, of course, nothing but a law of physics. Every sand particle has inertia, and the forces at work are gravity and friction. But this alone does not yet explain why avalanches occur. The behavior of sand piles can only be understood by looking at them in their entirety.

Let us first focus on sand at the micro level. Sprinkle grains of sand, one by one, on a table. The grains attempt to reach the lowest level of energy by seeking out the lowest-possible surface. If one carried out the experiment with a fluid, it would spread across the entire tabletop and, eventually, spill over the edge. Grains of sand, however, cling to each other because of friction. So as along as the pile is relatively low, every additional grain will remain on the very spot on which it was dropped. The landing place may happen to be the top of a grain, which in turn may lie on another grain. Once several layers of grain lie on top of each other, additional grains may slip over the side and, possibly, trigger an avalanche.

Small avalanches, but also big ones, ensure that the steepness of the sand pile does not exceed a certain limit. During experiments scientists observed that the strength of the sand avalanches was governed by a law that Beno Gutenberg and Charles Richter had proposed in the 1950s to describe the occurrence of earthquakes: Quakes of a magnitude of 6 occur 10 times more rarely than those that show a magnitude of 5 on the Richter scale.

The phenomenon of sand avalanches is an example of a so-called self-organized criticality. This phrase was coined by Per Bak, a Danish physicist, in 1988.* Together with Chao Tank and Kurt Wiesenfeld, Bak found that systems which consist of very many nearly identical components spontaneously find a certain state around which they will vary. In the case of sand piles it is the steepness of the flanks that determines the critical state.

Bak and his colleagues formulated their ideas on the subject wholly on a general basis. Hence, their conclusions are not limited to sand piles. Different systems or events such as forest fires, traffic jams, stock market crashes, and evolutionary processes are governed by the same laws.

Take the stock market. An investor, let's call her Mrs. Wall, could decide to sell her stock once its price reached a certain level. Mr. Street, a colleague who always follows what Mrs. Wall does, decides to follow suit and sell his shares. Others may follow the lead of Wall Street, and this may trigger selling orders of yet more investors. It is quite conceivable that the behavior of only a few investors will cause a selling wave and trigger a collapse of the market. Indeed, statisticians have observed that stock market crashes, some smaller, some bigger, occur at similar frequencies as do the avalanches that destroy the sand piles.

Yet another example of self-organized criticalities is traffic jams. Even before the relaxing vacation on the seaside, one has to brave the journey to the beach. Traffic is

---

*Bak died in October 2002 at age 54.

slow but steady. All of a sudden a driver in front of you hits the brakes. If the individual cars do not follow each other at too close a distance, nothing untoward will happen. But just as a single grain of sand may cause an avalanche, so a small braking maneuver by one driver can, if traffic is heavy and cars are dense, cause the dreaded traffic jam. Statistically speaking, the number of cars backed up in a traffic jam is comparable to the size of sand avalanches.

43

Buzzing Around Corners (Biology)

Can anyone claim not to have been driven crazy by the sound of buzzing flies? Fly swatters are of little help, since just about every time one tries to swat, the insect manages to avoid the threatening object by making split-second changes to its flight path. This is not surprising since flies require only 10 wing beats to make an acrobatic turn, lasting all of a 20th of a second. But how do flies perform these so-called saccades, these sudden turns in midair?

Two factors could conceivably affect the aerodynamics of a fly. One is the friction of the fly's skin in the air. The other factor that could play a role is the body's inertia, which keeps the flying fly on its course. For 30 years the assumption was that the aerodynamics of large animals, such as birds and bats, is determined by inertia. Flies, it was commonly thought, were too small for their inertia to have any significant impact. Scientists maintained instead that sudden changes in the flight direction of small animals were determined by the friction of their skin with the air. Flies swim in the air, as it were.

But Steven Fry from the Institute of Neuroinformatics, a joint institute of the Swiss Federal Institute of Technology and the University of Zurich, together with his colleagues Rosalyn Sayaman and Michael Dickinson from the California Institute of Technology, put an end to this—as it turns out—erroneous belief. In a paper published in the journal *Science,* they investigated the aerodynamic mechanisms underlying the free-flight maneuvers of fruit flies (*Drosophila melanogaster*).

The researchers set up three high-speed digital cameras in a specially equipped lab. At a speed of 5,000 images per second, each camera filmed the maneuvers of the flies as they approached, and avoided, an obstacle.

The recorded data were then loaded onto a computer-controlled mechanical robot. It consisted of artificial insect wings, constructed to scale and submersed in a tank filled with mineral oil. With the help of their robot fly, the three scientists were able to measure the aerodynamic forces generated by the wing motions of flying insects.

Their experiment yielded some remarkable observations. To initiate a saccade, the fruit fly creates torque through slight differences in the motions of its two wings. But the main point of interest was the fly's behavior after the onset of a turn. If indeed friction in the air were the determining factor when the fruit fly performs a saccade, a few beats of the wing would suffice to surmount the resistance. Then the fruit fly could quickly adjust its wings to their normal position and continue on with the flight ahead. But the researchers noticed something different. A split second after the onset of a turn, the fly produces a reverse torque with its wings for the duration of no more than a few wing beats.

Why does the fly do this? After initiating a turn—but already having stopped producing additional torque with its wings—inertia causes the fly to continue to rotate. Similar to a figure skater performing a pirouette, the fly continues spinning around its own axis. To counteract this continuing rotation of its body, the fly "puts on the brakes." Since this countersteering technique is only necessary to counteract inertia, the three researchers proved that it is inertia and not friction that is the determining factor when the fruit fly takes to the air.

44

Inexperienced Traders Make the Market Efficient (Economics)

During the very first lectures of Economics 101, students learn that, according to Adam Smith, supply and demand determine the price and quantity of goods sold. In reality, however, this relationship can only rarely be verified. Usually, markets are affected by numerous other forces that cannot be controlled and therefore falsify the results.

To make headway in understanding markets and economic behavior, economists initially resorted to additional assumptions, parameters, and variables. But the difficulties persisted, and instead of getting closer to a solution, all that happened was that the models became more intricate, unmanageable, and unrealistic.

So economists, taking a page from the books of physicists and chemists, tried to verify their theses by carrying out experiments. They set up labs in which "real" market situations were simulated as well as possible. Test persons were then observed making decisions and participating in economic activities. It was the birth of experimental economics.

Edward Chamberlain of Harvard University pioneered the experimental approach to economics some 50 years ago, using Harvard students as guinea pigs. Unfortunately, his results deviated from neoclassical market theory. The experiments showed volume to be typically higher and prices to be typically lower than was predicted by competitive models of equilibrium. A few years later Vernon Smith, one of Chamberlain's students, refined his former professor's methods. His experiments yielded near-equilibrium prices and quantities and finally gave experimental confirmation of the classical theory. For his work Smith received the Nobel Prize for Economics in 2002, together with the American-Israeli behavioral psychologist Daniel Kahneman, who pursued the same line of research.

In a new experiment the economist John List, from the University of Maryland, subjected the classical theory yet again to an experiment. In doing so, he made some remarkable observations that have been published in the *Proceedings of the National Academy of Sciences*. The economic goods that List used were baseball cards, hot items among aficionados. To find his guinea pigs, the professor went to a collectors market. He asked a number of traders and visitors if they would be willing to participate in an experiment. The "players" were then subdivided into four groups: buyers, sellers, novices, and experienced "old hats." Each seller received a card of a well-known baseball player. The image had been purposely defaced through the addition of a moustache to the player's likeness. Thus the card had lost all its value for real collectors, which ensured that the participants in the experiment would not take off and trade the card for money at the real market next door.

Then List allocated maximum buying prices to each buyer and minimum sales prices to each seller. These reserve prices were staggered in such a way that they produced offer and demand curves that intersected at seven cards and a price of between $13 and $14. For five minutes the participants were allowed to find partners and to higgle and haggle, dicker and chaffer, until they could agree on a price—or not. Efficiency of this artificial marketplace was measured in terms of whether the price and volume of the cards "sold" in the experiment corresponded to the predictions of classical theory.

Indeed, the results of List's experiment approximated the results predicted by theory. In 18 out of 20 cases between 6 and 8 cards were traded, and in 10 cases the average price of the cards equaled the predicted value.

But List noted a further detail: Market experience played an important role in determination of the market's efficiency. It was at its most efficient when inexperienced buyers chanced up experienced sellers. And the market was at its most inefficient when both buyers and sellers were experienced old hats. Apparently, too much familiarity with the rules of the game may prove a hindrance to trading partners, who are keen to find a price acceptable to both parties—a somewhat discouraging observation for believers in a free market economy.

45

The Waggle Dance of Internet Servers (Computer Science, Biology)

The Roman scholar and writer Marcus Terentius Varro (116–27 BC) was already convinced that bees were excellent construction engineers. Inspecting their hexagonal honeycombs, he had a suspicion that this structure was the most efficient in terms of enclosing the greatest amount of space for the storage of honey with the least amount of wax. But as recently demonstrated, bees could also be good computer engineers. Two millennia after Varro, Sunil Nakrani from Oxford University and Craig Tovey from the Georgia Institute of Technology presented a paper on mathematical models of social insects at a conference. Mimicking the behavior of honeybees foraging for nectar, they arrived at an efficient way for the optimal load distribution of Internet servers.

In the 1930s the Nobel Prize–winning zoologist Karl von Frisch discovered that the so-called waggle dance of bees inside the hive provides other bees with information about the distance and quality of a flower patch. Idle bees observe the waggle dance of one of their colleagues and set out on their run. (Since it is dark in the hive, they do not "see" the dance but infer it from changes in air pressure.) They do not communicate with each other before their flights, so none of them know how many will harvest nectar from which patches. And yet honeybees maximize the rate of their nectar collection. Beggarly flower patches are foraged by few bees while profitable and close-by patches receive a plethora of nectar-collecting bees. This happens due to so-called swarm intelligence: Even though each individual bee follows only a small set of instructions, the swarm in its entirety displays near-optimal behavior.

Sunil Nakrani and Craig Tovey were interested in the problems that Internet service providers face. An Internet service provider offers several Internet services—for example, auctions, stock trading, flight reservations. Based on the predicted demand for each service the provider allocates a certain number of servers (a cluster) to each service. The two scientists designed a model of server allocation based on the bees' behavior and ran simulations to test it. Incoming user requests are distributed into queues for the various services. For each completed request, the provider receives a payment. The number of incoming orders for each service continuously varies, and it would be profitable for underutilized servers to be allocated to overstretched clusters. This entails a cost, however, since a redistributed server must be reconfigured and loaded with the software for the new service. During that time—generally about five minutes—requests and orders cannot be met. If the waiting period (downtime) becomes too long, disappointed customers turn away and potential profits are lost. Hence, to maximize profits, providers must continuously juggle their computers between different applications and adapt to the changing levels of demand.

Traditionally, three algorithms have been used to calculate profitability. First, there is the "omniscient" algorithm, which determines at set time intervals the allocation that would have been optimal for the preceding time slice. Then there is the "greedy" algorithm, which follows the rule of thumb that the levels of demand for all services during a time slice remain unchanged during the next time slice. Finally, there is the "optimal-static" algorithm, which calculates—retrospectively—the optimal, unchanged (static) allocation of servers for the entire time period.

Nakrani and Tovey compared the honeybees' strategy to these three algorithms. In their model the request queues represent flower patches waiting to be harvested, individual servers represent the foraging bees, and server clusters represent the group of bees harvesting a particular flower patch. The waggle dance becomes a "notice board" in their model. After having fulfilled the requests, servers will, with a certain probability, post a note with the particulars of the served queue. Other servers read the posted notes with a probability that is greater the less profitable the queue is that they presently serve. On the basis of their own recent experience and the posted note, the servers decide—like worker bees observing a waggle dance—whether to switch to a new queue. The costs of a switch from one Web application to another one are comparable to a honeybee's time investment when observing a waggle dance and switching flower patches.

What the simulations showed is that in terms of profitability the behavior of bees collecting nectar outperforms two of the three algorithms by between 1 and 50 percent. Only the omniscient algorithm yielded higher profits. But this algorithm, which computes the absolutely highest upper limit of profitability, is not practical for two reasons. First, it is unrealistic to assume that future customer behavior is exactly known in advance. Second, the computer resources necessary to calculate the optimal allocation are enormous.

An aside: It was only in 1998 that the American mathematician Tom Hales was able to prove that the six-sided honeycomb (the hexagonal lattice) represents the most efficient partitioning of a plane into equal areas. (See Chapter 9.) But bees are not perfect, despite their apparent ability to approximate the optimal configuration in two dimensions. In three dimensions the much-lauded honeycombs are only close to optimal. The Hungarian mathematician László Fejes Toth designed honeycombs in 1964 that use 0.3 percent less wax than those built by the bees.

46

Turbulent Liquids and Stock Markets (Finance, Economics)

There is nothing surprising about the stock market going up or down on any given day, indeed at any given minute. Any economics student will have learned in the first weeks of his or her freshman year that the supply and demand of investors wishing to maximize their profits will affect share prices. The implicit assumption made is that traders react to all incoming information in a rational and reasonable way.

But then there are times when wholly unexpected fluctuations occur on the financial market that just cannot be explained by turning to "classical theory." A particularly dramatic market event took place on the London exchange on September 20, 2002. On this day, at 10:10 a.m., the FTSE 100 index rose from 3,860 to 4,060 within a span of five minutes. Within another few minutes the index fell to 3,755. After some further oscillations, which lasted for 20 minutes, the index returned to the initial value. This completely inexplicable spook saw some traders cashing in on hundreds of millions of pounds while others faced losses running into the same figures—and it all happened in less than half an hour.

Wild deviations, as observed in London, and the more generally observed regular oscillations remind observers, respectively, of turbulences in liquids and of muted vibrations produced by guitar strings. It is thus not surprising that physicists feel called on to search for explanations that could shed light on stock market behavior. Sorin Solomon from the Hebrew University in Jerusalem together with his student Lev Muchnik developed a model that offers some explanations for the puzzling events taking place on stock markets.

The difference between their model and the more conventional models is that the Israeli physicists did not assume that there exist classes of traders who differ from each other merely in terms of their reactions when faced with risk. What they did do was postulate different investor types. The interaction between these diverse participants in the stock market was too complicated, however, to be described by mathematical formulas. To get to the bottom of what actually happens on the stock market, they simply observed a simulated model over a period of time. This, they thought, was the way to get a firm grasp of what happens on the trading floor.

In Solomon and Muchnik's model there are investors who buy and sell based on whether the current price of shares is above or below market value. Then there are a few large investors who practically dominate the market and whose activities have a direct impact on share prices. Finally, the model also includes naive traders who base their buy and sell decisions simply on their past investments. In this model, which also considers other factors like the arrival of news items and various market mechanisms, virtual traders act autonomously. But their collective actions determine the behavior of the market.

After feeding all the variables into a computer, Salomon and Muchnik set up a simulation model for a virtual stock market. Can the existence of the three types of investors help explain the puzzling oscillations on the stock market? Lo and behold, their program produced behavior as it is observed in real-life stock markets. Damped oscillations occurred, suddenly interspersed with violent turbulences. Does that mean that real markets consist of these three types of traders? This is by no means certain, but what the model does at least show is how the interactions between different types of investors may lead to the surprising and sometimes colossal events that happen on the stock market and how the value of an investor's portfolio can change drastically in the time span of a coffee break.

47

Encrypting Messages with Candles and Hot Plates (Cryptography)

Whenever data are being transferred over the Internet, encryption methods keep PIN (personal identification number) codes secret, store medical information securely, watch over the confidentiality of online transactions, permit electronic voting, and verify digital signatures. In principle, cryptographic methods rely on the irreversibility—or at least on the difficult reversibility—of a mathematical operation. They depend on the fact that no algorithms exist that can invert a certain operation within a reasonable amount of time.

Operations that are solvable in one direction only are called one-way functions. A "one-way function with a trapdoor" is a mathematical function that is also solvable in the other direction but only if one possesses an additional piece of information—the encryption key. For example, it is a simple task to multiply two numbers. Dividing the result is much more difficult: Various candidates must be tested until a divisor is found that leaves no remainder after the division.

This is why the multiplication of prime numbers is nowadays used to encrypt messages. The intended recipient chooses two prime numbers, multiplies them, and publishes the product. Whoever wants to send her a secret piece of information will use this number to encrypt the message. The inverted operation—that is, the division of the product into its prime factors—is currently impossible if the number is large enough. Only a recipient who has the keys—who knows the two prime numbers—can decrypt the message. Thus the multiplication of large prime numbers is a one-way function with a trapdoor because the division of the product into its prime factors is impossible . . . unless one knows one of the factors.

Actually, nobody has ever been able to prove rigorously that it is impossible to factorize large numbers within a reasonable amount of time. And as ever-faster computers appear on the market, and ever-more-sophisticated algorithms are developed, searches for the appropriate keys become more efficient. These developments threaten the encryption methods that are currently in use. In 1970 the factorization of a 37-digit number was still a sensation. Today's world factorization record already stands at 160 digits: on April 1, 2003 (no April Fools' Day joke), five mathematicians at the German Federal Office for Security in Information Technology in Bonn managed to factorize such a number into its two 80-digit components. And no end to the developments is in sight. Could it be that the CIA, the MI5, or the Mossad already are in possession of an effective key search algorithm and just do not let on? In any case, for security reasons, numbers with at least 300 digits are currently recommended for encryption.

But a new technology—quantum computers—threatens to make numbers with 300 or even 3,000 digits vulnerable. In contrast to binary digits (bits) that can only take on the two states zero and one consecutively, quanta can simultaneously exist in more than one state. This means that, in principle, quantum computers could perform an enormous amount of mathematical operations simultaneously. Computations like the factorization of a large number, which would take centuries on a classical computer, could be performed within seconds.

For the time being quantum computers are still pie in the sky. Nevertheless, information technology officers, Web designers, and security professionals are on the lookout for encryption methods whose security is not a question of technology but would be guaranteed by the laws of nature. Recently two Swiss mathematicians suggested a method that may prove immune against attacks even by quantum computers. In a recent issue of the journal *Elemente der Mathematik*, which is devoted to articles about developments in mathematics in Switzerland, Norbert Hungerbühler of the University of Fribourg and Michael Struwe of the Federal Institute of Technology in Zurich presented an encryption method that relies on the second law of thermodynamics. This law, one of nature's most basic tenets, states that certain physical processes are not reversible. For example, while it is a simple task to prepare a café au lait—brew coffee, add milk, stir—it is quite impossible to separate café au lait into its components. Hence the preparation of a café au lait is a one-way function . . . without a trapdoor.

Another example of the second law is heat flow. Think of a hot plate under which several candles are lit. If the initial conditions—the locations of the candles—are known, the propagation of heat can be easily computed. On the other hand, according to the second law, it is quite impossible to trace the heat distribution back to its starting points; that is, you can never determine which parts of the hot plate have been heated by the candles. Even if the distribution of heat in a hot plate is well known at a certain moment in time, the initial positions of the candles cannot be deduced.

Hungerbühler and Struwe make use of this fact to suggest a novel "public key" encryption method. Let us say Alice wants to send an encrypted message to Bob. The two partners choose configurations of candles under their hot plates. These are their secret keys, α and β. Then Alice and Bob use the heat flow operator (H) to compute—each for him/herself—the heat distributions in the hot plates after, say, one minute (α*H and β*H). These heat distributions represent the public keys, and Alice and Bob publicize them in a directory or transmit them over public channels. Since the heat distribution can only be computed in the forward direction, a potential eavesdropper will find it impossible to deduce the initial positions of the candles from knowledge of the public keys.

Now Alice encodes her message, using her secret candle configuration and Bob's heat distribution (α*β*H). This is the heat distribution that occurs if both sets of candles are put under the hot plate. Because of the commutativity of the heat flow operator—it does not matter which set of candles is set up first—Bob can then decode the message with the help of his secret candle configuration and Alice's publicized heat distribution ($\beta * \alpha * H$). At the same time he can verify that Alice was the sender. The proposed encryption method does not depend on technology. It is based on the classical laws of nature and on the mathematical properties of thermodynamics. Therefore, it cannot be threatened by novel computing methods.

Unfortunately, it will not be possible to implement ThermoCrypto in the foreseeable future. The reason is that the equations describing the heat flow are continuous functions. Computing them with digital computers means truncating the numbers. The inevitable rounding errors could provide a starting point for eavesdroppers, and absolute security would no longer be guaranteed. Ironically, it is the quantum computers that could come to the rescue. In the mid-1980s the physicists Richard Feynman and David Deutsch had pointed out that due to its infinite states a quantum computer would be able to simulate continuous physical systems by making rounding errors infinitely small. So quantum computers, which may make conventional encryption methods obsolete one day, may also be the tool for the next generation of encryption methods.

48

Fighting for Survival (Evolutionary Theory, Finance)

When the mathematician John von Neumann and the economist Oskar Morgenstern wrote their groundbreaking classic on game theory at Princeton in the 1940s, they based their work on axioms—basic assumptions—which postulated that players are completely rational. The two theoreticians assumed that so-called *homo oeconomicus* (economic man) possessed all the information about his environment, was able to solve even the most complicated calculations in split seconds, and was not influenced in any way by personal preferences or prejudices. He would always make the mathematically correct decision.

A few years later the French economist Maurice Allais, Nobel laureate for economics in 1988, recognized that when answering questionnaires about situations that involved very low probabilities and very high sums of money, test persons tended to make "incorrect decisions." Their real-life decisions contradicted conventional expected utility theory. A few decades later Amos Tversy, from Stanford University, and Daniel Kahneman, from Princeton University, found that market participants—be they businesspeople, doctors, or just regular consumers—make decisions that contradict the axioms established by theoreticians in both regular and momentous situations. (Kahneman was awarded a Nobel Prize for Economics in 2002.)

The theoreticians were not easily put off by this contradiction between theory and reality. *Homines oeconomici* who did not behave according to the postulated axioms were simply labeled as irrational. The theory is correct, claimed the scientists. It must be that a large part of society simply reacts incorrectly. What these economists did not realize was that by stubbornly sticking to this belief they distanced themselves more and more from reality.

Herbert Simon, winner of the Nobel Prize for Economics in 1978, attempted to explain why financial markets conformed only poorly with people's behavior as predicted by game theory. He introduced the theory of "bounded rationality." By noting that agents must bear costs when acquiring information, must face uncertainties, and are not able to calculate like machines, he came closer to the truth. But this development was no cure-all either, and the anomalies observed in the financial market became more and more obvious. Above-average gains and losses were more common than classical theory would have led one to expect, volatility was larger than predicted, and exaggerated expectations led to inflated prices. Time and again market players who did not give a toss about the Von Neumann–Morgenstern axioms achieved higher profits than their more rational colleagues. Scientists had to look further afield for other explanations.

During the past century, economists often turned to other academic disciplines for tools to help them answer their questions about decision science and financial theory. A new fad among a young generation of financial theorists is evolutionary biology. Professors at renowned universities have made the theory of evolutionary finance an important topic for their research. Rumor has it that fund managers also make use of recent research results in this new field.

In early summer 2002 the Swiss Exchange invited scientists and practitioners from all over the world to a seminar in Zurich. Presenting their newest results, those present did not hesitate to criticize classical game theory for being removed from reality. While classical financial theory assumes that investors attempt to maximize their discounted income streams over the long run through clever investment strategies, evolutionary theorists suggest that investors follow just a few simple rules that they adapt from time to time to the varying circumstances.

In analogy to biological processes, economists set up

models of socioeconomic developments—complete with selection, mutation, and heredity—as a stream of learning processes and surges in innovation. In games that are rapidly played one after the other, the investment strategies assume the role of the animal species, and capital is allocated to the different strategies by the rules governing natural selection. Investment funds employing profitable strategies attract much capital and thrive, while funds with inferior investment strategies eventually disappear. Furthermore, surviving strategies must continuously adapt themselves to the changing market environment in accordance with the laws of natural selection.

The central question is, which investment strategies survive in an environment that is shaped by uncertainty and is prone to disasters? If several strategies operate side by side initially, which ones will survive in the long run? How do traders react to unexpected disturbances from outside? A paper by Alan Grafen of Oxford University may serve as an example for the approaches presented at the conference. In his model market players are equated to living organisms. To maximize their own level of fitness, they are subject to the process of natural selection and they adapt their behavior to the environment and to the strategies employed by competitors.

What Grafen observed was that agents will not get involved in any complicated calculations as they would in the classical theory. All they do is follow simple rules. If these rules prove successful and yield satisfactory results, their pervasiveness in the market increases. Under certain circumstances their dominance could, however, prove detrimental. Once weak strategies have disappeared, even successful strategies will no longer yield high returns, since nobody will be left to be exploited. Then the successful strategies will also disappear, just as predators become extinct once there is no prey for them to hunt anymore.

49

Insults Stink (Neurosciences, Economics)

A colleague is promised $10 on the condition that he share this amount with you. If you agree to accept the offer, both of you receive your shares; if not, both of you will get nothing. The colleague suggests sharing the amount equally. Would you accept?

Of course you would! And both you and your friend would go home well pleased, clutching $5 bills. But what if your colleague stops for a moment to reflect and realizes that, since he is the one making the offer, he could just as well keep $9.50 and offer you 50 cents. Would you accept? Most of us would refuse indignantly. "Who does he think he is? I'd rather do without half a dollar than let this scoundrel get away with $9.50 on my account!"

Reactions such as these have been replicated in experiments all over the world. This is somewhat surprising because they contradict traditional economic theory. After all, refusing even 50 cents is not rational. The offer of 50 cents may not be fair, but the alternative—going home with nothing—is worse still. Why then do people who find themselves in such situations behave irrationally?

For years this so-called ultimatum game has given economists headaches. They always assumed that economic decisions are firmly based on rational thought processes. A decision maker calculates the costs and benefits of his actions, weighs the probabilities of certain scenarios, and then makes the optimal decision. This is the fundamental assumption on which economic theory is based.

But what has emerged over the course of the years and been borne out by experiments with the ultimatum game is that while this assumption may well hold for decisions of institutions—firms, for example, and government agencies—it does not apply to decisions made by individuals. As it turns out, human beings do not make decisions based solely on hard facts and on the calculations of one's own advantage. They are also guided by emotional factors, such as envy, prejudice, altruism, spitefulness, and numerous other human weaknesses.

To explain the paradoxical results of the ultimatum game, scientists suggest evolutionary mechanisms. Refusing a derisory amount—so the argument goes—serves to uphold one's reputation. "I am not a wimp. Next time he will think twice before making such an insulting offer." Scientists believe that, in the long run, the social reputation of an individual may increase his survival chances.

Researchers at Princeton University and the University of Pittsburgh chose a different approach in the hope of gaining a better understanding of the decisions involved in the ultimatum game. They investigated the physiological processes that take place in the brain. This reductionist approach to the study of economic decisions—basing them purely on chemical and mechanical interactions between neurons, axons, synapses, and dendrites—is a novel way of conducting research into economics and decision theory.

The psychologists and psychiatrists who made up the research team subjected 19 test persons to the ultimatum game. The players, who had to compete against both humans and computers, were attached to magnetic resonance imaging scanners. These scanners highlight the regions in the brain in which changes in blood flow indicate increased activity of the nerve cells.

According to a report in the journal *Science*, their study was successful. They identified the regions in the brain that were activated during the ultimatum game. But somewhat to their surprise, not only the regions that are usually activated during thought processes—the prefrontal dorsolateral cortex—became busy. Rather, a region that is generally associated with negative emotions was also activated. And the more insulting the financial offer, the more intense the activity in the nerve cells became. The so-called anterior insula is the same area in the human brain that is activated in cases of strongly felt aversions, for example, when individuals are exposed to offensive smells or tastes. Insults stink.

But they had yet another surprise in store for them. The players' responses depended on whether it was an individual or a computer who made the offer. Unfair offers made by electronic calculators resulted in less activity in the anterior insula and were refused less frequently than unreasonable offers made by humans. After all, one does not let oneself be insulted by a computer.

50

Bible Codes: The *Not So* Final Report (Theology)

When the editors of the scholarly journal *Statistical Science* decided, in 1994, to publish the paper "Equidistant Letter Sequences in the Book of Genesis," they did not realize they were going to kick up a controversy that would last for more than 10 years. The question that the authors—Doron Witztum, Eliyahu Rips, and Yoav Rosenberg—investigated was whether secret messages were hidden in the Book of Genesis. The alleged messages pointed to events that took place only millennia after the writing of the Bible.

According to Jewish law, the Hebrew text of the Bible must not be changed even by a single letter in the course of myriad transcriptions. This is why, even today, the content of the Bible is believed by many to be identical to the text originally dictated to Moses by God on Mount Sinai.

The three authors believed they had found statistical proof for the existence of Bible codes. If the text of the Book of Genesis is strung along a line, without spaces, and letters are picked out at regular intervals, word combinations allegedly arise that make some sense. These words are called ELSs—equidistant letter sequences. (The intervals can be of any length, sometimes thousands of letters.) The *Proceedings of the National Academy of Sciences* had rejected the paper, but since the mathematical tools that were used seemed sound, *Statistical Science* agreed to publish it. The editorial board did not take the paper's claim all too seriously, however, and in an introduction to the piece questioned its scientific validity. The purported discovery of Bible codes was not referred to as a scientific achievement but as a puzzle.

Witztum, Rips, and Rosenberg claimed that the ELSs

of pairs of words that bear a relationship lie closer together in the Book of Genesis than would be expected by pure chance. To prove their thesis, they inspected the names and dates of birth and death of 66 famous rabbis. (In Hebrew, numbers are represented by combinations of letters.) To the authors' satisfaction, pairs of ELSs that referred to the same rabbi seemed to lie significantly closer than in randomized texts or when incorrect dates were assigned to the rabbis. This, they claimed, proved with high probability that the Bible predicted the appearance of Jewish scholars many centuries before they were born. A cryptographer at the National Security Agency, Harold Gans, followed up with an analysis of his own in which dates were replaced by the names of cities in which the scholars had been active. His study also seemed to show that the textual proximity of the ELS pairs was not solely due to chance.

The news that messages of the Almighty had been deciphered created a furor. In 1997 the first best-seller about the Bible codes appeared, and skeptics started to take notice. Brendan McKay, a mathematics professor from Australia, and Maya Bar-Hillel, Dror Bar-Natan, and Gil Kalai from Israel set out to debunk what they were convinced was pseudoscientific humbug. As was to be expected, the skeptics did not find any statistical evidence of hidden codes. Worse still, they charged that the data in the original paper had been "optimized," a euphemism for the charge that Witztum, Rips, and Rosenberg had adjusted the raw material to suit their investigation. After appraisal by several statisticians, their assessment was published in *Statistical Science* in 1999.

If the editors had thought the controversy would now be laid to rest, they were utterly mistaken. The debate was only heated up by the second paper. It did not dampen the enthusiasm of the proponents of Bible codes that "secret" messages were soon also identified in *Moby Dick* and *War and Peace*. In this charged atmosphere, scientists at the appropriately named Center for Rationality at the Hebrew University in Jerusalem decided it was time to subject the question of Bible codes to a sober, scientific analysis. A five-member commission was created and charged with getting to the bottom of the matter. It consisted of proponents of such codes and objectors and skeptics. Among the commission members were mathematicians of the highest caliber, like Robert Aumann (2005 Nobel Prize in economics), a leading mathematical expert in game theory, and Hillel Furstenberg, a world expert on an esoteric part of mathematics called ergodic theory.

Why does the test of the well-documented thesis turn out to be so difficult? One of the problems is that there are no vowels in Hebrew, and words appear much more often than in other languages, just by randomly juxtaposing letters. A randomly chosen collection of letters would spell out the name of a city, say Basle, with a probability of 1 to 12 million, while the same word in Hebrew, *Bsl*, would appear with a much higher probability of 1 to about 10,000. (There are only 22 letters in the Hebrew alphabet.) A more important reason for the controversy is that names can be written in different manners in Hebrew, especially when transcribed from Russian, Polish, or German. How, for example, should the German town where Rabbi Yehuda Ha-Hasid was active during the 12th century be written: Regensburg, Regenspurg, or Regenspurk? The resulting flexibility leaves researchers many degrees of freedom when preparing the database.

To eliminate any doubts about the data-gathering process, the commission charged independent experts with compilation of the place names. As a measure of precaution their identities were to remain anonymous and instructions were to be given in writing. After everything had been decided on and painstakingly recorded, the commission started its work—by throwing all instructions overboard. Experts were instructed partly in writing, partly orally, and part incorrectly. Some misunderstood the explanations; others made orthographic mistakes like confusing the Spanish towns Toledo and Tudela, Rabbis Sharabi and Shabazi, and places of death and of burial.

Then things started unraveling. Two members left the committee while it was still in its initial stages, and of the three remaining professors, one refused to sign the final protocol. Consequently, the majority report that was published in July 2004 was issued by just two members (Aumann and Furstenberg). Two others wrote minority reports, while the fifth had lost all interest in Bible codes and no longer wanted to be disturbed. That two out of five committee members do not form a majority is only one of the inconsistencies in the committee's work.

The "majority report" suggested that there was no statistical evidence of codes in the Book of Genesis—which is, of course, not equivalent to the statement that no Bible codes exist. The minority reports charged that the experiment was fraught with errors and therefore void of any meaning. In rejoinders, Aumann and Furstenberg rebutted the charges, and new statements followed. In the meantime the criticisms, refutations, responses, and rejoinders to the responses—all prepared with forensic meticulousness—fill binders. Words like liar, fake, and fraud that are rarely found in academic disputes were used by all sides. The three original authors publicly bet a million dollars that more ELS word pairs exist in the Book of Genesis than in Tolstoy's *War and Peace.* There were no takers, but Furstenberg challenged the proponents of the codes to design a test that would be more meaningful. Probably the most poignant statement was made by Aumann: No matter what the evidence, everybody would stick to his or her preconceived ideas.

References

7
TWINS, COUSINS, AND SEXY PRIMES

Dan Goldston and Cem Yildirim, "Small gaps between consecutive primes," http://aimath.org/primegaps/goldston_tech.

Andrew Granville and K. Soundararajan, "On the error in Goldston and Yildirimi's 'Small gaps between consecutive primes'," http://aimath.org/primegaps/residueerror/.

9
THE TILE LAYER'S EFFICIENCY PROBLEM

Tom Hales, "The honeycomb conjecture," *Discrete and Computational Geometry*, vol. 25 (2001), pp. 1–22.

10
THE CATALANIAN RABBI'S PROBLEM

Preda Mihailescu, "Primary units and a proof of Catalan's conjecture," *Journal für die reine and angewandte Mathematik*, forthcoming; "A class number free criterion for Catalan's conjecture," *Journal of Number Theory*, vol. 99 (2003), pp. 225–231.

13
HAS POINCARÉ'S CONJECTURE FINALLY BEEN SOLVED?

Grisha Perelman, "The entropy formula for the Ricci flow and its geometric applications," DG/0211159, "Ricci flow with surgery on three-manifolds," DG/0303109, and "Finite extinction time for the solution to the Ricci flow on certain three-manifolds," DG/0307245, http://arXiv.org.

18
GOD'S GIFT TO SCIENCE?

Stephen Wolfram, *A New Kind of Science*, Wolfram Media, Inc., Champaign, Ill., 2002.

23
KNOTS AND TANGLES WITH REAL ROPES

Yuanan Diao, "The lower bounds of the lengths of thick knots," *Journal of Knot Theory and Its Ramifications*, vol. 12 (2003), pp. 1–16.

Burkhard Polster, "What is the best way to lace your shoes?," *Nature*, vol. 420 (2002), p. 476.

25
IGNORANT GAMBLERS

"The communication of risk," *Journal of the Royal Statistical Society: Statistics in Society*, vol. 166 (2003), pp. 203–270.

26
TETRIS IS HARD

Erik D. Demaine, Susan Hohenberger, and David Liben-Nowell, "Tetris is hard, even to approximate," http://www.arxiv.org/PS_cache/cs/pdf/0210/0210020.pdf.

33
HOW CAN ONE BE SURE IT'S PRIME?

Manindra Agrawal, Nitin Saxena, and Neeraj Kayal, "PRIMES is in P," http://www.cse.iitk.ac.in/news/primality.html.

34
A MATHEMATICIAN JUDGES THE JUDGES

Lawrence Sirovich, "A pattern analysis of the second Rehnquist U.S. Supreme Court," *Proceedings of the National Academy of Sciences*, vol. 100 (2002), pp. 7432–7437.

37
COMPRESSING THE DIVINE COMEDY

D. Benedetto, E. Caglioti, and V. Loreto, "Language trees and zipping," *Physical Review Letters*, vol. 88 (2002), p. 48702.

38
NATURE'S FUNDAMENTAL FORMULA

Johan Gielis, "A generic geometric transformation that unifies a wide range of natural and abstract shapes," *American Journal of Botany*, vol. 90 (2003), pp. 333–338.

40
THE FRACTAL DIMENSION OF ISRAEL'S SECURITY BARRIER

B. B. Mandelbrot, "How long is the coast of Great Britain? Statistical self-similarity and fractal dimension," *Science*, vol. 155 (1967), pp. 636–638.

G. Dieter and Y-C. Zhang, "Fractal aspects of the Swiss landscape," *Physica A*, vol. 191 (1992), pp. 213–219.

41
CALCULATED IN COLD ICE

Naohisa Ogawa and Yoshinori Furukawa, "Surface instabilities and icicles," *Physical Review E*, vol. 66 (2002), p. 41202.

43
BUZZING AROUND CORNERS

S. Fry, R. Sayaman, and M. H. Dickinson, "The aerodynamics of free-flight maneuvers in *Drosophila*," *Science*, vol. 300 (2003), pp. 495–498.

44
INEXPERIENCED TRADERS MAKE THE MARKET EFFICIENT

John List, "Testing neoclassical competitive market theory in the field," *Proceedings of the National Academy of Sciences*, vol. 99 (2002), pp. 15827–15830.

45
THE WAGGLE DANCE OF INTERNET SERVERS

Sunil Nakrani and Craig Tovey, "On honey bees and dynamic allocation in an Internet server colony," *Proceedings of 2nd International Workshop of Mathematics and Algorithms of Social Insects*, Atlanta, Ga., 2003.

47
ENCRYPTING MESSAGES WITH CANDLES AND HOT PLATES

Norbert Hungerbühler and Michael Struwe, "A one-way function from thermodynamics and applications to cryptography," *Elemente der Mathematik*, vol. 58 (2003), pp. 1–16.

49
INSULTS STINK

Alan G. Sanfey, James K. Killing, Jessica A. Aronson, Leigh E. Nystrom, and Jonathan D. Cohen, "The neural basis of economic decision-making in the ultimatum game," *Science*, vol. 300 (2003), pp. 1755–1758.

50
BIBLE CODES: THE *NOT SO* FINAL REPORT

Robert Aumann and Hillel Furstenberg, "Findings of the Committee to Investigate the Gans-Inbal Results on Equidistant Letter Sequences in Genesis," Center for Rationality, Discussion Paper 364, June 2004, http://www.ratio.huji.ac.il/dp.asp.

Index

D

I

J

K

L

M

N

O

P

S